Does an electron have a structure?
— *read the unexpected answer...*

*****

ÅKE HEDBERG

Cosmic rays.

The content

The chaos theory **14**

The problem of emergence **17**

Natural philosophy **21**

God and the brutal mathematics **24**

Mathematics by its nature is inextricably linked to a mechanical logic **25**

The physics and cosmology **26**

Metaphysical elements into physics **27**

The natural dialectical particle theory **28**

Photon dimensions and its components **30**

Here's how protons and anti-protons are formed **36**

How the W-boson is composed **40**

Four different values of time **43**

The electron charge q **43**

What causes the electron mass? **45**

How does a muon really decide when and how to decay? **48**

The forces that hold the electron components **50**

Postscript **54**

///

© Åke Hedberg 2017
Förlag: BoD – Books on Demand, Stockholm, Sverige
Tryck: BoD – Books on Demand, Norderstedt, Tyskland
ISBN: 978-91-7699-661-4

DOES AN ELECTRON HAVE A STRUCTURE?
– read the unexpected answer…

✳ ✳ ✳

The electron is probably the most famous of all particles in physics and the most investigated. It was also the first of the kind to be discovered. On this and many other things about the electron type Wikipedia[1]:

> In the Standard Model of particle physics, electrons belong to the group of subatomic particles called leptons, which are believed to be fundamental or elementary particles. Electrons have the lowest mass of any charged lepton (or electrically charged particle of any type) and belong to the first-generation of fundamental particles. The second and third generation contain charged leptons, the muon and the tau, which are identical to the electron in charge, spin and interactions, but are more massive. Leptons differ from the other basic constituent of matter, the quarks, by their lack of strong interaction. All members of the lepton group are fermions, because they all have half-odd integer spin; the electron has spin 1/2.
>
> Electrons play an essential role in numerous physical phenomena, such as electricity, magnetism, and thermal conductivity, and they also participate in gravitational, electromagnetic and weak interactions. Since an electron has charge, it has a surrounding electric field, and if that electron is moving relative to an observer it will generate a magnetic field. Electromagnetic fields produced from other sources (not those self-produced) will affect the motion of an electron according to the Lorentz force law. Electrons radiate or absorb energy in the form of photons when they are accelerated. Laboratory instruments are capable of trapping individual electrons as well as electron plasma by the use of electromagnetic fields. Special telescopes can detect electron plasma in outer space. Electrons are involved in many applications such as electronics, welding, cathode ray tubes, electron microscopes, radiation therapy, lasers, gaseous ionization detectors and particle accelerators.
> ///..//
> The electron has *no known substructure and it is assumed to be a point particle with a point charge and no spatial extent*. In classical physics, the angular momentum and magnetic moment of an object depend upon its physical dimensions. Hence, the concept of a dimensionless electron possessing these properties contrasts to experimental observations in Penning traps which point to finite non-zero radius of the electron. A possible explanation of this paradoxical situation is given below in the "Virtual particles" subsection by taking into consideration the Foldy-Wouthuysen transformation.
>
> The issue of the radius of the electron is a challenging problem of the modern theoretical physics. The admission of the hypothesis of a finite radius of the electron is incompatible to the premises of the theory of relativity. On the other hand, a point-like electron (zero radius) generates serious mathematical difficulties due to the self-energy of the electron tending to infinity.
>
> Observation of a single electron in a Penning trap suggests the upper limit of the particle's radius to be

---

[1] It is not due to laziness I quote from Wikipedia; the reason is rather to provide mainstream, the commonly accepted and so to say official version of an approach, a theory, etc.

10^{-22} meters. The upper bound of the electron radius of 10^{-18} meters can be derived using the uncertainty relation in energy.

There is also a physical constant called the "classical electron radius," with the much larger value of 2.8179×10^{-15} m, greater than the radius of the proton. However, the terminology comes from a simplistic calculation that ignores the effects of quantum mechanics; in reality, the so-called classical electron radius has little to do with the true fundamental structure of the electron. (my italics).

**

In this essay, I will primarily focus on the question of the electron has a clear structure. Or if what is said above in the quote from the Wikipedia applies, that is to say: *The electron has no known substructure and it is assumed to be a point particle with a point charge and no spatial extent.*

And this: *The issue of the radius of the electron is a challenging problem of the modern theoretical physics. The admission of the hypothesis of a finite radius of the electron is incompatible to the premises of the theory of relativity. On the other hand, a point-like electron (zero radius) generates serious mathematical difficulties due to the self-energy of the electron tending to infinity.*

Let's follow a physics forum[2] in physics that just discusses the question whether the electron has a substructure. The time is January 2009.

Qubix asks:
> *Does an electron have an internal structure?* Ok, I know this question is very old, and it has probably been answered by now, but if the electron does not have an internal structure (like the proton for example), how does it maintain itself as an entity? Why does it not disintegrate? I've asked this question a couple of times before ,and people answered with things like "it's a fundamental particle so it does not have an internal structure"... this is obviously not a scientific answer, for 100 years ago, we might have said the same about the atom.

Mupp answers:
> What's not scientific about it? Science is the investigation of the physical world by means of experiment in order to advance our understanding of nature such that we not only know why we obtain the results of the experiments already performed, but are able to predict the outcome of future experiments (or explain why a certain experiment is intrinsically impossible to perform, or why a certain prediction is intrinsically impossible to make, etc.) Sometimes an experiment proves a well-established theory wrong, and that's when science is perhaps most interesting. But at other times you have to accept that the scientific method is intrinsically incapable of proving something beyond absolutely all doubt, and you just have to accept that a theory which continually churns out the right answers is probably at least a very good approximation to what's going on, an approximation that is so good you might as well call it correct for expedience's sake until something proves it wrong. The theories that are currently the best we have predict that the electron is a structureless, fundamental particle with zero size. It doesn't disintegrate because there is nothing for it to disintegrate into, and nothing to "maintain". At some point, it is almost logically necessary for such a fundamental particle to exist, or you'd have an infinite regress-regress-howhow would the particles which made up the electron maintain their structural integrity? We know that atoms aren't fundamental for lots of reasons, but perhaps the most obvious piece of evidence that they have structure is that we can smash them into bits. You can't do this with an electron. People have tried investigating the size of the electron, which if it were non-zero would tell us that our present best theories are wrong. So far the upper bound on any potential size of

---

[2] https://www.physicsforums.com/threads/does-an-electron-have-an-internal-structure.287118/

the electron is (I think) about $10^{\wedge -18}$ m — or a one-hundred millionth the size of the atom.

Qubix, muppet said:

What's not scientific about it? Saying something is "fundamental" is the same as saying:"it was made by god". It is not a scientific definition because if you do not have a theory capable of understanding a phenomenon, it does not mean that the phenomenon itself is the problem, but most likely, your theory. What if I asked another question: What is the electric charge? The purpose of the question is to go beyond (or around) the "its fundamental" lame answer, and to see if modern theories (even hypotheses like string theory), can say something about it. People have tried investigating the size of the electron, which if it were non-zero would tell us that our present best theories are wrong. So far the upper bound on any potential size of the electron is (I think) about 10^{-18} m — or a one-hundred millionth the size of the atom. In the first sentence, you are implying the size of the electron is 0? Size 0 would mean the electron is a point. Can you really have a point in a (at least) 3 D space?

malawi_glenn (Science Advisor Homework Helper):

Yes, it is a point. The thing is that our ideas of the world in our daily life contradicts what is going on in the Quantum level. Also the problem might lay in the mathematical language we use, which was developed to explain classical physics (Newton , etc.) But you know that geometric series etc converge to a finite value, even though each terms goes to zero but never becomes zero? That also contradicts how we add things in our daily life, but in the world of math it is just fine. So if we want to be really pragmatic scientific, the answer is that our theories for electrons (The Standard Model) has a delta-function as the electrons charge distribution and that quantity can be tested experimentally — the form factor should then equal unity. And that is what is found in all experiments so far, the upper limit for the electron radius is around 10^{-18} m. So that is the most scientific answer you can get today.

DrChinese (Science Advisor Gold Member):

Qubix said: Saying something is "fundamental" is the same as saying: "it was made by god". It is not a scientific definition because if you do not have a theory capable of understanding a phenomenon, it does not mean that the phenomenon itself is the problem, but most likely, your theory... In the first sentence, you are implying the size of the electron is 0? Size 0 would mean the electron is a point. Can you really have a point in a (at least) 3 D space? First, welcome to PhysicsForums, Qubix. Electrons have mass and charge even though they act as point particles. Same thing is true of quarks, the other major building blocks of matter. However, electrons and quarks can act as either a particle or a wave according to how they are observed. Saying a particle is fundamental is not at all the same as saying it was made by god. The theory says that an electron is fundamental (i.e., does not desintegreits to other particles like free neutrons do, for instance). It does not do so as far as anyone knows (and people have looked). Theory also says an electron is a point particle, and it acts like a point particle. So yes, you can have a point in spacetime. (There are a lot of that are seen in physics that are counter-intuitive, no point in denying what is known to occur.) Experiment again matches theory. What do YOU think an electron is? How does your concept differ from accepted theory? And do you have any experimental basis for your opinion?

Algr answers:

Is it perhaps better to say "We can't see any evidence that the electron has an internal structure." ? I take it then that we have never seen an electron decay or turn into anything else without something impacting it first. BTW, protons and neutrons are made of quarks — are there any particles that have electrons in them? Or are electrons always solitary? Jan 25, 2009

malawi_glenn answers:
> DrChinese, we should not perhaps encourage doubters to do wild speculations here? Algr, the atom has electrons. You would then say that an atom is not a particle, but then I would not call the proton a particle. It is all about energy scales here. And to the OP, suppose THAT the electrons was made up of smaller particles, or strings, we would then ask the question "what are those particles made of?", so we have to accept that a smallest entity exist I think.

DX answers:
> We don't know whether the electron has internal structure. If it does have internal structure, it is not accessible at the energy levels that we can currently probe.

AEM:
> So if an electron has no internal structure and for all intents and purposes can be regarded as a point particle, what is the origin of its mass?

ZapperZ (Staff Emeritus Science Advisor Education Advisor Insights Author 2016 Award AEM) said:
> So if an electron has no internal structure and for all intents and purposes can be regarded as a point particle, what is the origin of its mass? If the LHC finds the symmetry breaking in the electroweak sector, then that's your possible source of leptonic mass. Zz.

Qubix:
> DrChinese said: What do YOU think an electron is? How does your concept differ from accepted theory? And do you have any experimental basis for your opinion? Well, besides what the accepted theory tells us, I do not know what the electron is, that was the whole purpose of my question :) Thank you for all your answers, I hope we find out more about the electron in the future.

AEM:
> ZapperZ said: If the LHC finds the symmetry breaking in the electroweak sector, then that's your possible source of leptonic mass. Zz. Can anyone cite a reference that will elaborate on this a little?

humanino friend:
> Qubix said:Saying something is "fundamental" is the same as saying "it was made by god". It is not a scientific definition because if you do not have a theory capable of understanding a phenomenon, it does not mean that the phenomenon itself is the problem, but most likely, your theory. I think your question is the same as what reason do we have to believe that the electron is fundamental. You don't trust experimental results because experiment can only rule out some theories, but never prove a theory since we might not be doing the right experiment (not high enough energy, or not done the experiment enough times, etc.) Others here can correct me if I'm wrong or expound on this idea. But IIRC, particles emerge from symmetries of spacetime, you know SU(3)XSU(2)XU(1). And it seems that the electron has all the properties of one or some of these properties which indicate that it is fundamental. If this is true, then it is interesting to consider what particles say about spacetime itself. And I'd have to wonder how quantum gravity theories would change the description of particles.

ZapperZ (Staff Emeritus Science Advisor Education Advisor Insights Author 2016 Award):
> AEM said: Can anyone cite a reference that will elaborate on this a little? http://arxiv.org/PS_cache/arxiv/pdf/0704/0704.2232v2.pdf Zz.

AEM:
> Thanks. Precisely what I was looking for.

humanino friend said:
> Particles emerge from symmetries of spacetime, you know SU(3)XSU(2)XU(1) This is not spacetime symmetry, this is gauge symmetry.

Algr: malawi_glenn said: Algr, the atom has electrons. You would then say that an atom is not a particle, but then I would not call the proton a particle. It is all about energy scales here. Well, that defines some terms, but doesn't answer the question at all. Is there anything smaller then an atom that has an election as a component?

malawi_glenn Algr said:
> Well that defines some terms, but doesn't answer the question at all. Is there anything smaller then an atom that has an election as a component? Well it was due to a sloppy usage of the word "particle" that I raised that issue. No, there are no other composite particles which are made up of electrons.

Algr said:
> Is there anything smaller then an atom that has an election as a component? The size of the atom is determined by the strength of electromagnetism. By which interaction would your thing be bound by ? Electrons do not feel the strong force.

malawi_glenn:
> Well, one can also have positronium, but if one can classify that as smaller than an atom, I don't know.

We do not need to ask more questions about the electron to understand that the theory of this our very first subatomic particle is mildly incomplete. There are too many who ask questions about the electron's internal structure, And the only answer given is a nonsense about points. Even such as the issue of the radius of the electron is a challenging problem of the modern theoretical physics. Or that:"People have tried investigating the size of the electron, which if it were non-zero would tell us that our present best theories are wrong."

And: "if an electron has no internal structure and for all intents and purposes can be regarded as a point particle, what is the origin of its mass?" And what if an "*electron has no known substructure and it is assumed to be a point particle with a point charge and no spatial extent,*" as someone in Wikipedia says, what kind of physics is it then?

One might ask that if this our most fundamental particle is so poorly understood, how about the other elements of the atom, such as the proton, the neutron and all the others as pioneer, myons, etc. Yes, what happen then if we ask about the quantized value of electric charge, the electron's elemental charge? Is it always the same? The question arose, and came to a head as a result of the investigation of the proton structure in the 1960s. It meant that the proton was composed of three different particles with third-party charges.

The somewhat unexpected result gave birth to the idea of new unknown particles: **quarks**. *Instead of asking deeper questions about and making more accurate investigations into the nature of the electrical charges*, the chosen road was chosen, and on a quasi-philosophical basis, to invent new particles as an explanation. But I will return to this, now with a series of very well-known experiments, reactions and decay processes, the electron is not actually a dot-shaped device but is composed of a pair of our most known particles and so that it really has an inner structure. This composition and structure also explain the electrical element quantum and its mass in a very convincing and surprising manner.

If I now claim that the electron has a clear structure when it is composed of other particles — for a long time known (not quarks) - it means that it is possible to "hit them in pieces" and thus their constituents are found so that It may also mean that this has already happened but not understood and therefore not observed. (I return to such an imaginable experiment of "splitting" electrons or "smashing them into pieces"). It may also mean that this experimental method is not applicable here, but also many others that prove its composition of other particles. We can study what happens in different forms of so-called decomposition of particles. We can then start with the most famous decomposition of a particle, known as the beta decomposition. (From Wikipedia, the free encyclopedia).

In nuclear physics, beta decay (β-decay) is a type of radioactive decay in which a beta ray (fast energetic electron or positron), and a neutrino are emitted from an atomic nucleus. For example, beta decay of a neutron transforms it into a proton by the emission of an electron, or conversely a proton is converted into a neutron by the emission of a positron (positron emission), thus changing the nuclide type. Neither the beta particle nor its associated neutrino exist within the nucleus prior to beta decay, but are created in the decay process. By this process, unstable atoms obtain a more stable ratio of protons to neutrons. The probability of a nuclide decaying due to beta and other forms of decay is determined by it's binding energy. The binding energies of all existing nuclides form what is called the nuclear valley of stability.

Beta decay is a consequence of the weak force, which is characterized by relatively lengthy decay times. Nucleons are composed of up or down quarks, and the weak force allows a quark to change type by the exchange of a W boson and the creation of an electron/antineutrino or positron/neutrino pair. For example, a neutron, composed of two down quarks and an up quark, decays to a proton composed of a down quark and two up quarks. Decay times for many nuclides that are subject to beta decay can be thousands of years.

Electron capture is sometimes included as a type of beta decay, because the basic nuclear process, mediated by the weak force, is the same. In electron capture, an inner atomic electron is captured by a proton in the nucleus, transforming it into a neutron, and an electron neutrino is released.

Outside the nucleus, free neutrons are unstable and have a mean lifetime of 881.5 ± 1.5 s (about 14 minutes, 42 seconds); therefore the half-life for this process./.../ Beta decay of the neutron, described above, can be denoted as follows:

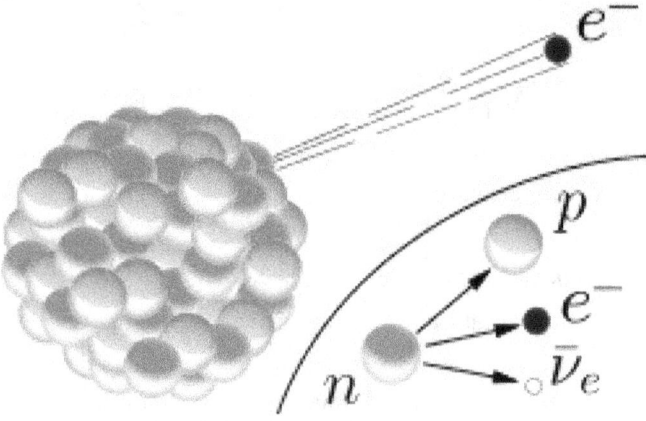

A schematic of the nucleus of an atom indicating β− radiation, the emission of a fast electron from the nucleus (the accompanying antineutrino is omitted). In the Rutherford model for the nucleus, red spheres were protons with positive charge and blue spheres were protons tightly bound to an electron with no net charge.

The inset shows beta decay of a free neutron as it is understood today; an electron and antineutrino are created in this process.

In both processes, the intermediate emission of a virtual W^- −boson (which then decays to electron and antineutrino) is shown down.

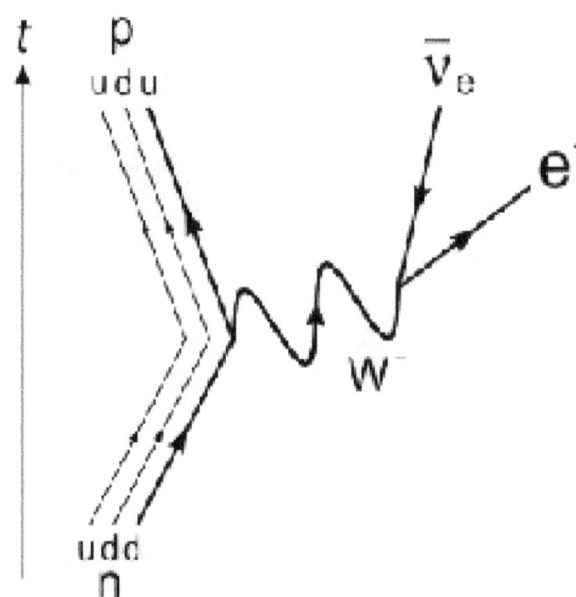

$$n^0 \rightarrow p^+ + e^- + \nu_e$$

where p^+, e^-, and ν_e denote the proton, electron and electron antineutrino, respectively.

For the free neutron the decay energy for this process (based on the masses of the neutron, proton, and electron) is 0.782343 MeV. The maximal energy of the beta decay electron (in the process wherein the neutrino receives a vanishingly small amount of kinetic energy) has been measured at 0.782 ± .013 MeV. The latter number is not well-enough measured to determine the comparatively tiny rest mass of the neutrino (which must in theory be subtracted from the maximal electron kinetic energy) as well as neutrino mass is constrained by many other methods.

A small fraction (about one in 1000) of free neutrons decay with the same products, but add an extra particle in the form of an emitted gamma ray:

$$n \rightarrow p^+ + e^- + \nu + \gamma$$

This gamma ray may be thought of as a sort of "internal bremsstrahlung" that arises as the emitted beta particle interacts with the charge of the proton in an electromagnetic way. Internal bremsstrahlung gamma ray production is also a minor feature of beta decays of bound neutrons (as discussed below).

The transformation of a free proton to a neutron (plus a positron and a neutrino) is energetically impossible, since a free neutron has a greater mass than a free proton.

Well, there is more text to download from Wikipedia, but we are pleased with this now. But first. As I see the decay process above:

$$\text{step I: } n^0 \rightarrow p^+ + w^-$$
$$\text{step II: } w^- \rightarrow e^- + \nu_e$$

I think the process will develop in two steps, not the only on as the common, conventional approach. I will discuss this problem later.

Before we look at the electron closer to its possible seams, we need to examine the light photos closer. Many believe that for this fundamental particle there is nothing more to add beyond what has been said a long time ago. But they are wrong. First, we need to figure out and answer the question Einstein asked more than 60 years ago:

> All these fifty years of conscious brooding have brought me no nearer to the answer to the question, 'What are light quanta?' Nowadays every Tom, Dick and Harry thinks he knows it, but he is mistaken. (Albert Einstein, 1954)

Yes, what are light quanta? My new particle theory is based on a new theory, also a new geometric model about how the photons work

To me, a photon is neither a wave nor a particle; there is a third: An oscillator, an LC circuit, reminiscent of a CD or a wheel.

The graphics above, with six photons and a simplified picture, represents an electromagnetic gamma ray. Moving with the velocity of light c. Below we see a single photon, a quantum of light seen from the front. Further explanation will come

and, therefore, what a light quantum is. The theory, in my opinion, answers Einstein's question and gives us the most fundamental building blocks of nature. For, from this light photon, the electron and proton plus many other particles can be derived in one way or another. The common theory has no art.

It may sound like a supernatural statement or a pure lie but is actually completely true, as is proven here. But then we have to go to the basics. On the purchase, a number of problems can be investigated and explained. Of course, I cannot come here with a complete answer to the question of light quanta, it requires a whole book. But on the limited space, a beam of light looks like this as shown in the figure below. It is thus composed of a number of photons, greatly simplified here and seen a little bit from the side (here five), which travels forward at the speed of light c.

If we consider a single photon which, coming towards us, it looks like this and with the help of a conventional / Cartesian coordinate system usually describe the geometry in this way:

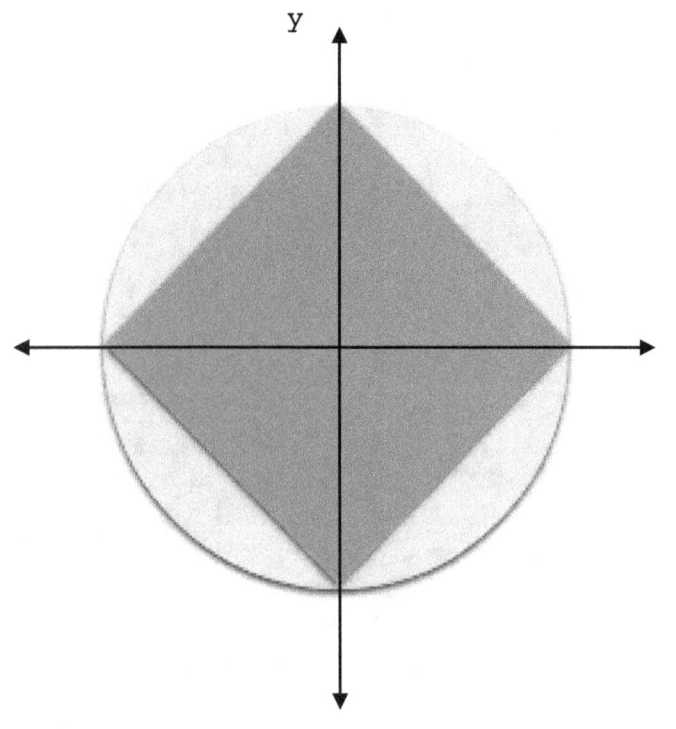

Here is the radius of the (yellow) circle **r** where the surface becomes **π * r²** while the square's surface (blue) is **2t²** when its diagonal is **2t**. But let's put below — what we might call — a pair of force arrows. (± F). Plus a red dot as the center / origo.

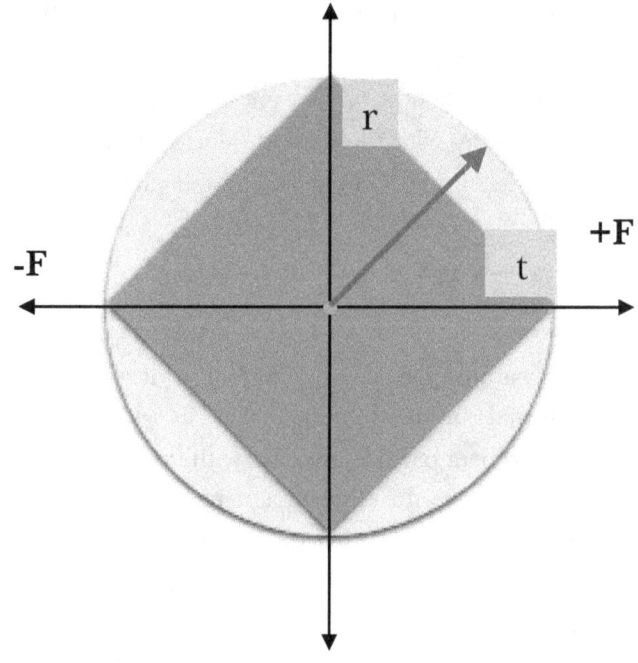

In next figure above to the right hand, Y is thus exchanged for r and t, as well as X. While a coordinate system as above is four-dimensional, the former is only two-dimensional. A crucial difference.

Another important difference is that the original system has no zero point (the red dot), it is a small four-dimensional surface, with the dimensions lp (= Planck-length), and tp, (= Planck-time). R is thus the maximum length and T maximum time. Mathematically, it means that:

$$\pi R * lp * 2T * tp = \pi r^2 * 2t^2 = h$$

Little h stands for h Planck constant. Welcome to the description of a new world – *the real world!* Unlike the Cartesian artifact as the ancient ruined, completely mechanical "world" that the common physics deals with. (A world which of course constantly held as enigmatic, filled with black holes, quarks , etc., it therefore appear to be. The world I discuss, analyze and compose it all hangs together. As we will see a little further in the text. It sounds like a floskel, but of course it is natural. In my world).

In this way, this coordinate system is now defined, which is not the norm when defining means delimiting. (The coordinate axes of the current system are unlimited long and the origins have the coordinates 0.0, what this means.) My coordinate system is thus more general, since each axis, horizontal as vertical, has the same dimensions (r and t). Time measured in seconds and length measured in meters *has the same weight and none of them is favored*, which does not apply to the normal geometry, where the dimension *meter* only applies. It therefore carries a system that mimics nature more than the usual one, which is a description of a bedroom, a kitchen or a laboratory. Only in rare cases time is introduced, but in a non logical way.

This, of course, also provides a different geometry. I believe the problem with the parallel axiom is solved if we define — delimit — the geometry according to my suggestion. A geometry according to nature. (A non-artificial, non-Euclid geometry.)

Wikipedia:

Parallel postulate:

The parallel axiom is the fifth axiom of euclidean geometry (named after the Greek mathematician Euclid). The axiom is more controversial than the other axioms because it is not as easy to formulate and the meaning is not considered by everyone to be as obvious as one often wants an axiom to be. Euclid tried to vain themselves to prove parallel axiom with the other four reasons.

If the parallel axiom is rejected and the axiom is replaced with another, non-Euclidean geometries are available. These new geometries are different theories and a certain rate can be true in one theory and one in another.

There are different formulas of the parallel axiom, but this (Play-fair's axiom) is probably the most common:

Given a straight line and a point beyond the line, one can draw one and only one straight line passing through the point and parallel to the line The parallel axiom is equivalent to the claim that the angular sum in a flat triangle is 180 degrees. (From Wikipedia)

In nature, therefore, there are no lines drawn endlessly parallel. Simply why there are no such endless lines, just finite. All distances and all times are finite. Ever since Galilee's days, it has been assumed that mathematics is the language of nature. But this is not the case. It is only in the eternal world of Absolute Mathematics and Religion, we find endless distances and times. Nature is determining, not abstract thoughtless mathematics.

The discovery of the Planck length (l_p) and the Planck time (t_p) given by natural measured constants gives us the evidence. Where the logic also requires that there is something small, the biggest must be. In this case, large R and T, the maximal distances and times. Then we will not be too surprised that the of these smallest and largest units times 2π becomes the physics perhaps most important constant, Planck's constant h.

This image of a new world could suffice to describe a new electron theory regarding an internal structure. But something is missing to complete the image of this new world. Namely, just a literally new world, a world beyond the normal. A world that mainstream physics touched upon but not been able to formulate a theory.

—

I could start by asking why only the horizontal dimension has the designation F (plus or minus). What about the vertical? Are they power less? Maybe friend of order and logic asks.

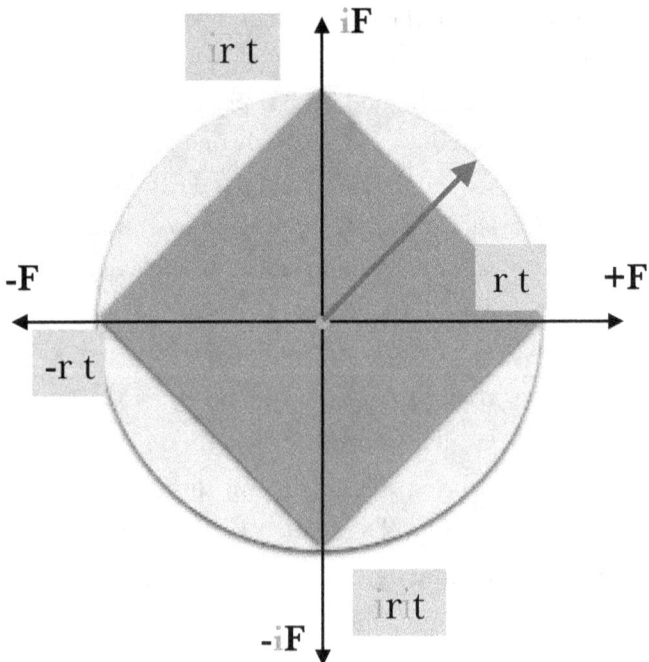

Can generalization to run longer? Yes, it does and therefore we definitely leave the normal theory behind us and at the very least the common view of science. We leave a one-sided, only mechanical world and enter into the now complete and real world. See above where all the "sensors" and "antennas" of the photon are printed. (Although a physicist of the old tribe would rather call them fields and field lines.) In any case, there is an effect and impact on the environment that the photon has, a power effect if desired. This force F is therefore always the same, a quantum equal to:

$$F = \sqrt{h/2\pi} = 1{,}02 * 10^{-17} \text{ newton.}$$

Of course, in more complicated cases, there are multiples of F, but for each individual photo regardless of size and frequency, F is always the same.

The letter F stands for Newton's name of power. In my view, it may therefore have four different forms as we see here. Plus F and minus F - effect and counter activity — is no problem, but what is ± iF for? The letter in stands for the so-called imaginary device, which can be described according to the following relationship where its square is negative:

$$i^2 = -1$$

It is not entirely wrong to see the different F as a power field with universal extent, and as a kind of antennas or so which feel and affect the environment, constantly maintain contact with, retrieve and deliver information from and to the outside world.

As we understand, the photon is an existential form of non-mechanical character. It's what we can say of electromagnetic nature that a physicist has to say, *but now something new*: The photon is an electromagnetic circuit with an imaginary existence form that a mathematician could say, a vibrating event containing a certain amount of motion equal to Planck's constant h Thus, the photons exist in a non-mechanical world, a kind of parallel world that is imaginary. A world not possible to depict. This world was almost discovered in the 1960s (and even closer to the 1980s) by the so-called chaos theorists when they designed chaos theory with the help of computers. More about the dimensions of the photos on page x and onwards.

The chaos theory

Within physics, the world of the photons or states, as previously mentioned, different names as electrodynamic or electromagnetic, in mathematics imaginary and non-real. Physically, it is about the aggregation state that is *above or beyond the three we usually think of: the solid, the liquid and the gaseous*. It is thus not an intentional condition or merely thought: it is an experimental very well-known state and exists in our world and reality but we do not notice it. It is about the chaotic state and its theory.

The chaos theory is the most important of the new theories in the latter part of the twenty-first century. In order to understand this, the imaginary device and complex numbers plus state description are required in graphical forms using computers. But it's not chaos theory itself, it's important, it's its integration with science and cosmology and by this the creation of a third new. This is precisely what motivates the subheading of new insights and discoveries with a new understanding of physics and cosmology in this writing. It is therefore an integration, not a mere addition. Many have already tried, but have not been able to say life and depth in the old theory formation.

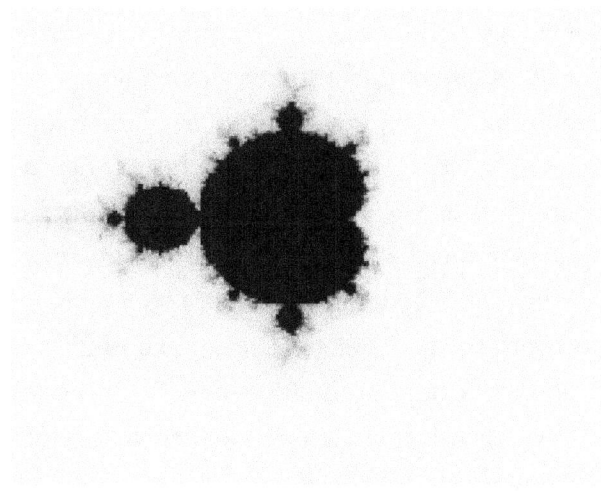

Fractal graphics. The process is described by iteration of complex numbers. Chao is an extreme aggregation state. A state I call absolute (not 100% Absolute).

Chaos theory, which certainly has roots far back in time, is thus the youngest of the major natural sciences. What has previously prevented exploration of this area has been the lack of advanced calculators. The exploration of complex numbers led Frenchman Gaston Julia and then Benoit Mandelbrot to the concept of the Julia amount and the Mandelbrot amount and the concept of fractals. With the help of computer graphics, these could then be linked with the new chaos and chaos theory. No tools have had such a big role to make the imaginary states as real as the computers.

The chaos theory, dynamic system theory, the study of nonlinear systems or complex adaptive systems, as it is also called, has something of a halo of mystery around. This may be regarded as a multidisciplinary research area, but does not really fit into any classical subject, as most other scientific sciences do. However, if one considers chaos as one of nature's aggregation states, the most extreme after the more well-known solid, liquid and gaseous states, it is undoubtedly a branch of physics.

It is thus a matter of nature's non-material side, its spiritual content. Once upon a time, about a hundred years ago, the physicists worked with concepts like the ether and different variants of this in attempting to fill this knowledge theory gap and to try to answer the question of what kind of medium the light actually spread. The concept Eater abolished Albert Einstein at the beginning of 1900 and then unfortunately replaced with the even more vague concept of "void". The question of what the light and all other electromagnetic radiation actually propagate in is strangely not yet answered, which is symptomatic of the mechanics and superficial philosophy of modern physics and its present absence of critical thinking.

It is thus primarily mathematicians who call the world or state we are talking about imaginary and non-real, in contrast to the real state or world. But they do not really talk about worlds or states but about numbers, imaginary and complex. In this writing, however, it is a description of a real world, a non-mechanical world that the existence of the complex numbers gives a hint of. However, many have indeed already touched on this in mind and also described and named this world without taking the plunge and realize its existence. James Clerk Maxwell, for example, developed the concept into a special light ether. This non-mechanistic world, however, belongs to our reality and can therefore be regarded as a side of it. For of course, our entire world and reality has, like a coin, two sides. Without this side our world is not complete, then it is at best only be called the half.

Many philosophers and mathematicians have thus turned to this non-real and imaginary world earlier but have not been able or dared to take the step and the thought fully. For example, in addition to Maxwell such as Leibniz and Gauss, as we will see in the forthcoming text. Quantum physicists have partially dealt with this imaginary side of the world, which is clear in their mathematical formalism, but they do not seem to fully understand what they were doing.

$$i\hbar \frac{\partial \Psi}{\partial t} = -\frac{\hbar^2}{2m}\frac{\partial^2 \Psi}{\partial x^2} + V(x)\Psi(x,t) \equiv \hat{H}\Psi(x,t).$$

```
Here we see (ih) how the imaginary
device is used in Schrödinger's equation.
But only as a mathematical emergency
aid ...this as only one of many examples.
```

In modern times, the British radio physicist Dr Paul J. Nahin[3], Professor Emeritus, has thoroughly dealt with the imaginary device. He has written several books on imaginary greats and its history, both in science and mathematics, but never taken the full step. He does not seem to dare or realize that it is often a real world he writes and tells about hundreds of pages. I have also written to him and asked about this but unfortunately no response.

Nahin has not understood Einstein once understood, namely how to interpret Max Planck's new constant h. He understood that there was a physique behind the numbers h stood for, the constant was not just a mathematical piece of art that Planck considered it to be. Einstein understood the new phenomenon of the photon and its value expressed in frequency times h, because he saw it as a physical reality (although confusing one) behind the numbers. Bohr disagreed because it made arbitrary choice of mathematical solution, he thought.

Priests and poets have also spoken and written about this non-material world at all times. Called the spiritual, divine, transcendent, etc. But when they had dubious intentions and been metaphysical, they could not further develop the idea, not fully accepted it and / or could have given it a scientific or mathematical meaning. They have neither been able nor dared. People in general have thought and talked about the world and reality in all ages at all times. Directly or indirectly. Of course. It is proven by both ancient cave paintings and literary texts. But for that reason, no theories have been formulated about their nature of a scientific nature.

But there is more that physicists know but choose to ignore. I am referring to something that everyone with natural science education knows and is particularly important in chemistry, but which in one way or another applies to all areas. I could say that the phenomenon is particularly important for understanding physics, but the truth is, unfortunately, instead, it is completely ignored. It's about the emergence. In all scientific

---

[3] Paul J. Nahin. Professor i electrical engineering på the University of New Hampshire. The Radio of Science, Springer-Verlag New York, Inc. 2001.

sciences, but also in economic, political and social areas, so-called emergence and catalytic effects are important; Sometimes completely visible and obvious, but not always. For some practical and pedagogical reasons distinguished educators at once upon a time physics and chemistry. Limits were established for what was theoretically applicable at one or the other level. For some physicists, chemistry was just something that smelled bad.

Okay, before moving on, I just point out that, depending on frequency, photons have a little different geometry. If r/t equals c, then the surface πr^2 of the circle is equal to $2t^2$ of the square. The square is thus enrolled in the circle, as it is called. Unless this similarity applies, they are different, but it is still stated that their is the same, that is, Planck constant h.

Emergence are such processes where a complex pattern is formed based on interplay between simple structures or behaviors.

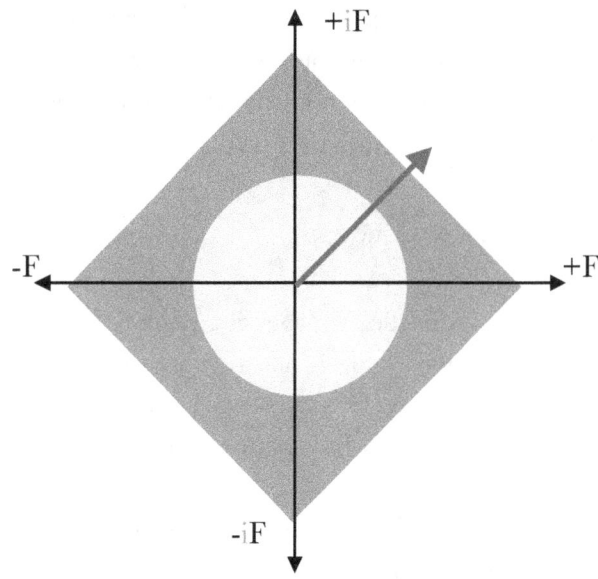

Above a photon with high energy and frequency.

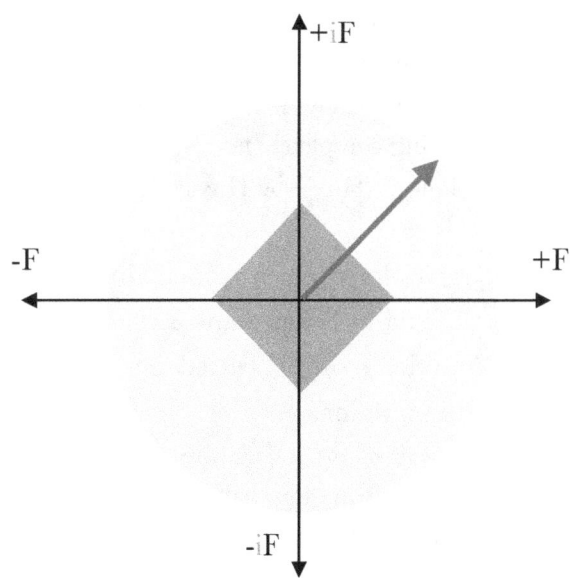

Above a photon with low energy and frequency.

This is what the graphics above and to the right are intended to describe.

Now to the problem of emergence.

But first let's look at a few lines about the emergence quoted from Wikipedia:

/ ... / Emergent structures are patterns not created by an individual event. Nothing commands a system to shape a pattern, rather, the interaction of several parts creates a complex chain that leads to a certain order. It can be argued that emergent structures are more than its combined parts, because an emergent order is not solely due to the mere existence of the parts – the interaction of the parts is central. Emergent structures can exist in many naturally occurring phenomena, in different subjects such as physics and biology. For example, the form of weather phenomena like hurricanes is an emergent structure. (Wikipedia).

And a few quotes:

1: Aristotle, Metaphysics, Book 8.6.1045a: 8-10: "... the whole is not as it was, but a gathering, everything is something beyond the parts ...", that is, the whole is greater than The sum of the parts.
2 "The chemical combination of two substances, as is well known, produces a third substance with properties distinct from those of the two separate substances, or the two substances together" (Mill 1843)
3 ^ Julian Huxley: "Once again, it's a sudden transition to a whole new and more comprehensive type of order or organization, with completely new emergent qualities, and involving completely new methods of continued evolution" (Huxley & Huxley 1947) (Wikipedia) .

Then we look on a course the "Emergent Phenomena in Science and Everyday Life", on an American site.[4] Created by: University of California, Irvine.

About this course: Before the advent of quantum mechanics in the early 20th century, most scientists believed that it should be possible to predict the behavior of any object in the universe simply by understanding the behavior of its constituent parts. For instance, if one could write down the equations of motion for every atom in a system, it should be possible to solve those equations (with the aid of a sufficiently large computing device) and make accurate predictions about that system's future.

However, there are some systems that defy this notion. Consider a living cell, which consists mostly of carbon, hydrogen, and oxygen along with other trace elements. We can study these components individually without ever imagining how combining them in just the right way can lead to something as complex and wonderful as a living organism! Thus, we can consider life to be an emergent property of what is essentially an accumulation of constituent parts that are somehow organized in a very precise way.

This course lets you explore the concept of emergence using examples from materials science, mathematics, biology, physics, and neuroscience to illustrate how ordinary components when brought together can collectively yield unexpected, surprising behaviors.

Note: The fractal image (Sierpinkski Triangle) depicted on the course home page was generated by a software application called XaoS 3.4, which is distributed by the Free Software Foundation under a GNU General Public License.

Upon completing this course, you will be able to:

1. Explain the difference in assumptions between an emergent versus reductive approach to science.
2. Explain why the reductivist approach is understood by many to be inadequate as a means of describing and predicting complex systems.
3. Describe how the length scale used to examine a phenomenon can contribute to how you analyze and understand it.
4. Explain why the search for general principles that explain emergent phenomena make them an active locus of scientific investigation.
5. Discuss examples of emergent phenomena and explain why they are classified as emergent.

---

[4] https://www.coursera.org/learn/emergent-phenomena

And so on. Ok, here are some examples. By itself, sodium (Na) and chlorine (Cl) are highly toxic and dangerous elements, the heavy green gas chlorine, for example, used as a fighter gas during the First World War. Do you eat a pea bit Na, you die a painful death. But in a interaction, together and chemically linked to sodium chloride - NaCl - they form something brand new and get a qualitatively new feature, namely the rather harmless table salt to keep on the eggs. This new feature is a result of the emergence. In military science speaks of binary war gases. It is such chemical weapons that consist of two separate, individually harmless gases, which when mixed immediately give rise to a dangerous . These are also examples of the emergence, which is abundant in chemistry, but also at higher levels. One swallow does not make summer ...

Perhaps the most extreme example of the emergence, however, is the combining of an egg and a sperm cell. Individually, they have no future, and soon they die: but united, they form a new developable cell that can rapidly form new cells and then a whole organism with its own life and a new future with a huge potential ability.

The most extreme example elsewhere, so to say, at a deeper level than the chemical and biological, we also find important emerging effects. Also on the purely physical, subatomic level. For example, have you understood neutrinon and photon, so you can understand what mass and electrical charge is. For neither the single unbound neutrinon nor the photon has any mass or charge. Just because these are emergent properties. Emergens also emerges here in association, together and in interaction with other particles – thus with other photons and/or neutrinos.

Thus, an individual free neutrino can not have the mass or charge property. But along with, associated with a photon, the emergence may occur so that the property of mass and electrical charge occurs. For example, what we know as electrons. Or when three neutrins (one of which is an antineutrino) form a proton.

The same principle applies to the photons. Two photons in interaction may cause the mass and charge property to occur, such as the $W^{\pm}$-particle and the Higgins particle. One of the photons can be an anti-photon. More about this will come later.

The fact that some researchers believe they have discovered that neutrino has a mass and has been awarded a Nobel Prize for this may be due to the fact that they have occasionally encountered a contact – and thus a certain emergence effect – with other particles during the experiments and measurements. Which caused the experiments to be misinterpreted...

—

Quantum mechanics (QM; also known as quantum physics or quantum theory), including quantum field theory, is a branch of physics which is the fundamental theory of nature at the small scales and energy levels of atoms and subatomic particles.

Classical physics (the physics existing before quantum mechanics) derives from quantum mechanics as an approximation valid only at large (macroscopic) scales.

Quantum mechanics differs from classical physics in that: energy, momentum and other quantities are often restricted to discrete values (quantization), objects have characteristics of both particles and waves (i.e. wave particle duality), and there are limits to the precision with which quantities can be known (uncertainty principle). (Wikipedia)

The theory we seek can paradoxically be considered as both global and basic and that the objects we are looking for are a special system, a structured whole composed of certain units we can call quantum. Thus, devices that can be decomposed and merged together form new devices with new features. I have initially mentioned the phenomenon in question and it is called emergence. Such we know from chemistry and biology, but now we discuss the matter at a more basic physical level, but also apply to a more global macroscopic — cosmic — level, which can thus be considered paradoxical.[5] More on this with the emergence comes later.

Someone may mean that on the physical level, the subatomic level, we have the *quarks* that fill that function. But on closer inspection, this is wrong. Quarks are artificial creations, based on introduced "objects" and phenomena that try to make sense of certain observations, such as third-party charges. On the other hand, they are not at the macrocosmic level we are referring to, they are merely mechanical objects to their intended nature. Though electrically charged. They are too many: together with their anti-forms, they are twelve and tend to grow. They also cannot explain photons, neutrinos and electrons. They, plus many "adhesive particles" – gluons – must also be added. All in all, the number of "basic" units becomes 30-40. *The theory of quarks cannot explain anything* and is a deadly dead end.

As can be seen from the quote from Wikipedia above, our view of quantum theory also differs from the view of the quantum concept. "Quantum mechanics /.../ is the fundamental theory of nature at the small scales and energy levels of atoms and subatomic particles", they say. In our case, it is not a question of a theory on one or the other level, it is a matter of *how our whole world is structured and composed*. Neither is it a matter of *mechanics* alone! The phrase "quantum mechanics" – quanta that are merely *mechanical* are revealing how briefly the entire mindset is. And the quantization would expire after a certain level is a poor but unconfirmed guess. Strange!

I find it even weird and absurd to equate the quantum theory of quantum **mechanics**, wave **mechanics**, matrix **mechanics**, etc. Is it that everything is just *mechanical*? Okay, you are also talking about quantum electrodynamics and quantum chromodynamics. As if nature allows itself to be divided in this way! Of course you can talk about nature mechanical or electrodynamic side, but that's another matter.

With these findings about the state of modern science, I have to say that there is no knowledge and understanding of the structure of nature and its way of working and, at all, to function. It seems that physicists do not get any further on the road they once hit. The electron becomes with its point shape and its electrical charge as a small impenetrable bullet which can not be further explored. It is as if the present science has entered a maze they can not find out or in a swamp that they cling around and unable to get a stop in.

They do not seem to have a clue about, for example, the theory of knowledge – the issues that employed The old Greek and Chinese minds very much. A fundamental but simple and obvious philosophy. Instead, clinging around in a ridiculous mechanical and unilateral world. (See, for example, Plato's view of the matter a little further on in the text). You can also think of the modern quantum field theory function when you do not even know what a quanta or a field is. A theory which sounds very, very sophisticated – always for a few, of course. For others one confused chatter.

―――

---

[5] See my book: *Universe without Big Bang & atoms without quarks*. Available on most online bookshops.

Natural philosophy

As always, man has tried to understand and philosophize how the world is built, functioning, and once emerged. An attempt has been made to get an understanding of the context and based on its constituents and foundations. Traditionally, the wise Thales, who appeared at the beginning of the sixth century before our counting, was considered in the thriving maritime and trading town of Miletos on the west coast of Asia Minor, as the founder of Western Natural Philosophy. His thesis was: "all things originator is water." The word "originator" here stands for the Greek word "arkhe" readable "principle" in the sense this word has in Aristotle which among other things had a general theory of the four elements plus an ether. Previous theories about elements before Aristotle developed primarily as a response to the radical critique of the cosmological theories especially Parmenides – the first knowledge theorist – had raised.

This had suggested that the ultimate thought object and its context must be a uniform, unchangeable being. A non-being could therefore not even be imagined. It was completely impossible, was his logic. Based on this, he also considered the problem of whether something could change at all. According to his logic, he concluded that there was no movement at all! End of discussion; all movement was impossible and everything was in fact a hoax a bluff, an illusion! What was available was "the One" a sort of spherical uniform motionless sphere.

His strict logic about the nature of movement at all is interesting to follow. A logical string that, in parentheses said in Platon's eyes, made him "reverent" but also "terrible" as this logic would have devastating consequences if it were true. It thus led to a contradiction – a paradox – which can be said gave rise to the axis around which the pre-Socratic philosophy came to circulate. Parmenide's student Zenon "proved" about the same as his teacher with his famous paradoxes about the Achilles and the Turtle, the Flying Arrow with several strict logical reasoning.

It is not entirely wrong to claim that the issue of the nature of light and its relation to an aristotelian ether has then played a crucial role throughout the natural science revolution, a revolution that can be said to have culminated in quantum mechanics in the 1920s and now come To end a long time ago.

The question of an ether, however, is still haunting, though it has been given other names and attributes. In the 1950s, the famous quantum mechanic P.A.M. Dirac[6] that although Einstein's relativity principle of 1905 led to the abolition of the ether with the new quantum electrode dynamics, we are still "rather forced to have an aether." The reason that the etheric concept lived and actually still lives (we are still talking about ether media when we are talking to radio and television, for example), I think has to do with the fact that the logical stringency requires something beyond the light itself to propagate In, a medium or the like.

Directly abolishing the ether, as Einstein did at the beginning of the last century, does not solve the problem and the difficult seemingly insoluble paradoxes, which an analysis of the nature of light easily leads to, as we shall see. The danger is, however, that, like Parmenides, ending up with some kind of "terrible" even more paradoxical logic. But what cares the modern physicist about this? Nada.

Thus, Parmenides resonated. He (who incidentally seemed to be the first to discover that the evening star and the morning star were one and the same, Venus) spoke of two kinds of philosophy. One was the truth philosophy and the philosophy of appearance. "Between two

---

[6] Nature 168: 906-7

ways the choice should meet: is and is not". And if that is not, nothing can be said or even thought. It does have "neither name nor thought". Remains the other, that is, and is real. If you travel the natural philosopher and natural scientist Thales Nature's law is water, Parmenides emphasized that Nature's law is that it is. The non-existent, which is not, by definition, was non-existent and, therefore, was both unthinkable and not possible to proclaim, he meant.

Parmenides polemizes also to the claim that the current could grow out of the non-being. He speaks perhaps therefore the Pythagoreans who were actually followers of non-being, and thought that the finite Cosmos thus received its nourishment from the surrounding void. But also to Anaximandros thesis about the ultimate ground and also to his vision of the Cosmos emerged from Chaos and his view of non-being and Apeiron, who were without quality and had a character of voids.

But here, Anaximandros and others who speak of an ultimate ground are guilty of a mistake, Parmenides said. You cannot talk about a being – the ultimate ground – and a non-being. It is not a consistent and logical philosophy. Either one or the other must apply and for Parmenides, the logic required that the ultimate ground is that it is. A non-existent was not intelligible, it was incomprehensible and did not even make it possible to name. In the name of the consequence, neither that which *is* neither come nor perish, neither may it be unchanged. The being must therefore be unnecessary and unavoidable. Yes, even more, it must be immobile and endless; No movement in real meaning cannot be found. Nor can it be said that something has been, when it still is. Yet another consequence of his team is that it must be one and everywhere equal, which follows that it is coherent and continuous. Although his strict logical reasoning is not so easy to review, it obviously leads to an unreasonability – a "terrible" paradox – about the real nature of the world and of life. But yet, it is not fully logical and does not seem to be in harmony with the common sense to say that if what is not, what by definition does not exist cannot be said much? If anything?

In any case, Parmenide's student Zenon who walked in his master's footsteps could also "prove" this logic. However, Zenon was an independent thinker whose thoughts are still discussed. He has been called by *the inventors of dialectics*, and by various evidence has sought to show that all arguments against Parmenide's teachings were incorrect. Based on that reality was intelligible, he had among other things, concluded that this was motionless and unlimited and therefore no extension in space. There were theories that Zenon would confirm and do it by showing that both the movement, the diversity and the room is paradoxical concepts. He meant and wanted to prove that reality to be intelligible not include such things as movement, diversity and space. For these phenomena are contradictory and such paradoxes cannot be with the real world to do. Then it would not be understandable. And it must be.

To the most familiar of these paradoxes are the Akilles and the turtle. Here Zenon proves that the fleet-footed Achilles never catches up with the tortoise provided the turtle get a certain advantage. For every time Akilles arrives at the animal's starting point, it's a bit longer. And that's true. Every time Akilles arrives at the place where the tortoise was recently, this has already left it. And although this is repeated for infinity, this is what Zenon said. Thus, Akilles never catches the turtle. Which should be proved. Another of Zenon's paradoxes is what is called the dichotomy or bisection paradox. Aristotle summarizes this: "Movement is impossible, because an object in motion must reach halfway before reaching the goal." For because there are so many "halfways", which requires endless time, the goal will never be reached.

Another of Zenon's famous paradoxes is the "flying arrow". It usually sounds like this: If a flying arrow moves, it must either move in the place where it is located or in the place where it is not. The latter is impossible. But the former is also impossible, for if the arrow is in a fixed location, so it does not move. Thus, the arrow does not move but stands still.

However, there were others who did not agree with this. In addition to the Pythagoreans also the Chinese thinkers, who argued that even non-being exist, although in a different and opposite shape than the real world. Even non-being was thus a form of existence which they formulated the yin and yang thinking and Anaximandros in his thesis on the origin of the world from the quality of loose and formless Chaos. But this qualityless is still "something", it owns both motion and activity. *Chaos is not passive and ineffective but highly active.* This approach is consistent with how the ancient Greeks and Anaximandros looked at what he called the Apeiron. This active base and primal ground he imagined to be both unlimited and quality-free. It was not endless, it was unlimited, which is a big difference.

For the adoption of a primal ground, Aristotle argues that there is an endless re-use of 'material' in order for this to be a constant innovation of things – a justification that may have been said by Anaximenes and possibly by the Pythagoreans, but which seems unnecessary in Anaximandro's system, when this counted on a cyclical shift of origin and fall. I also count "with a cyclical shift of origin and fall."

What about the "quality-free" primal ground, it's also interesting to note that "Anaximandro's system" is also true to my view on the whole. Since the apeiron must be the primal ground and in its modifications to show all qualities in nature, that is, when it has to possess all qualities in nuce itself, it must be quality-less, because the opposite qualities equate each other. We must consider this as the logical motivation for the quality of the primal ground.

An interesting and logical motivation. Not least, the perception that this quality-less root cause itself must possess all (smoothed) qualities in order to "display all qualities in nature." Everything we find in our common everyday world will thus be traced back to Chaos! This is where the modern information concept is so fruitful. The source of all information – all in this our common world – has its true origin in the (yet) unformed information in Chaos. Or how to express it.

What is meant by the term "quality"? Qualities such as green, red, salty, sour, etc. have their way of being - that is not the thing without properties, or ways of being, says Aristotle. Quality in physical sense can also be a movement of some direction and size. Then it is not a matter of things but a movement with certain characteristics. If such a motion exists, an exact same motive motion is eliminated, and the sum becomes zero and without quality. In other words, Newton's law of action and reaction applies Anaximandros systems. Or differently expressed: if each movement is reflected in both size, shape and direction, it means that the "opposite qualities equalize each other" and it is both quality-less and formless.

—

This is therefore important to remember: every assumption of a non-being as something non-existent, something that does not belong to our world and existence leads to "terrible" paradoxes; to absurd contradictions, which ultimately literally lead into an impossibly empty nothing. We have a logic that would have devastating consequences if it were true, in the words of Plato. A first lesson we can draw from this is that it is not enough (apparent) logic, we must learn and try to understand how nature actually works. Ask nature questions and then try to figure out a logical answer. It is also the philosophy I here and now tries to apply.

But it is precisely this "terrible logic" that Plato warns that permeates throughout the so-called modern physics and cosmology. Do not wonder that it's packed with so many unresolved riddles and mysteries!

God and the brutal mathematics

Modern "solutions" to these paradoxes are not lacking. All they do have in common that they are of mathematical nature. The philosophical ontological, knowledge-theoretical solutions are completely lacking. For example, the Finnish-speaking philosopher Erik Stenius writes in his book *The Dawn of Thought* the following as he claims is the solution of paradoxes:

> The reasoning by which the Zenon wanted to show that Achilles not be making the turtle is based on the premise that the sum of an infinite number of time distances are always endless. Since this premise follows an incorrect conclusion, it must be incorrect – belief that it applies constitutes an error of thought. If you do not make this mistake, there is no contradiction in the Zenonian reasoning. So if there is no better reason to explain the motion unintelligible, there is no reason whatsoever.

And Stenius adds:
> To this end, someone might face: But I think the amount of infinite number of time intervals must always be endless. And then a mathematician answers if he's brutal: in that case, you must learn to realize that you think wrong. No other option is available.

The "brutal" dictatorship, the God father of mathematics, has spoken. Stop thinking, do not lift your eyes from the desk, do not look through the window, do not study reality and nature yourself! Some mathematician's premise is that in this our real world there is endless time and space. That the infinite and even "many infinites" are a reality. But I would say that this applies only in the world of "pure" mathematics and religion, where the Absolute, Infinite and perfect God to 100 percent rule. Otherwise, it is this way some physicists invent black holes, dark matter and energy, inflation theories, etc. as they blindly follow the God father mathematics instead of thinking for themselves, hopping for a while and following some kind of reason.

However, this is how a swedish physicist[7] of the old tribe writes, one who understood to distinguish mathematics and physics:

> But as the mathematician can count on zero and infinity in a fully consistent and logical manner, should the physicist also be able to pronounce on infinitely distant things, provided he, like the mathematician, knows the laws and knows how to apply them? Then you've been thinking so long. The simple beauty of Newtonian mechanics in conjunction with its unprofitable practical successes seemed hypnotic. One imagined the universe as a celestial machinery, working with mathematical precision. /.../ The physics has gone through a crisis whose suites have not yet been completely overcome ./...// But there is a significant difference, from a theoretical point of view, between mathematical statements and statements about nature. The mathematical contexts are based on a foundation of aprioristic axioms. Through logical proof, deduction, increasingly complex mathematical conclusions are compiled. However, no new knowledge is added. /.../ But physics is something completely different from mathematics. Although mathematics plays an important role in physics, there is a gap between mathematical theories and physical

---

[7] Tor Ragnar Gerholm, *Fysiken och människan*. Stockholm 1963

theories. Ever since Galileo and Kepler, the physicists have consistently sought to formulate their natural laws into mathematical symbols.

But emphasizes TRG:
> Physics is not a kind of mathematics: Physics's thesis are, of course, quite different from mathematics. They are not analytical without synthetic. Synthetic thesis express themselves about experimental results and observations and the conclusions that can be drawn from them. They are not "empty" of actual content, but add something new to our knowledge. We get to know something that we did not know before, something that was not implicit in the premises.

"All knowledge of reality," Einstein says, "based on experience and derives from it. On purely logical terms, thesis won are completely ineffective with regard to the real content." Unlike the world of mathematics and religion, and when it comes to "statements about nature," one cannot count "zero and infinity." We have no experience or knowledge about this. Infinite time, endless space is not in the reality I'm talking about. Here is the principle that the absolute is also and can be relative, and the relative is and can and will be absolute.

As the Polish-Jewish doctor and the philosopher Ludwik Fleck (1896-1961), the Absolute (in capital letters) must be relativized to absolute and absolute the relative. The relative is relatively relative to something, thus to the absolute. And vice versa. All other talks become pure nonsense, not intelligible. The statement, for example, that "everything is relative" means a contradiction, a paradox, since the claim itself must be an exception to this rule, directly contradicting it, since it is of Absolute nature – not even absolute in small letters.

Our world is not Absolute and unlimited or infinite, there is no endless time and no endless space. Neither is it Relative for the same reason. In addition, the absolute and the relative (lowercase) are always in a context. And this context is nature and reality. Not in the perfect and divine books of religion or in the high spheres of pure mathematics. *The solution of paradoxes lies in this.*

Mathematics by its very nature is inextricably linked to a mechanical logic...

The Swedish mathematician Anders Karlqvist asks questions about the nature of mathematics, which is unusual. As a mathematician, he surprised the mathematical nature and wonder if it can be mathematical errors, it *cannot express anything other than the mechanical* processes and procedures. He asks:
> Mathematics, in fact, by its very nature is inextricably linked to a mechanical logic, where the logical rules like gear drives the mechanical "solutions". Is there an alternative or is it that the human world is not accessible via mathematical thinking?

The common mathematic therefore is unable to express *anything but mechanical things*. ((Unless we execute operations with imaginary and complex devices, though stuck in the mathematical sphere without contact with the world in general). Treating the world as if it only has a mechanical side also binds it to eternal thinking, to its Absolute nature, that is, a religious thought. And then, of course, it will not be "accessible", as it does not allow itself to be defined. Believing that mathematics as such "is inextricably linked to a mechanical logic" can of course not be completely correct. As indicated above, it has an another and more democratic side, something that will soon be dealt with.

On the other hand, the kind of mathematics that today's physicist applies is merely a mechanical nature. And then quantum mechanics, mathematical mechanics and all modern physics are called *nothing but a set of mathematical formulas,* (many have never understood this but think it has some kind of physical logic) every statement it comes with must be questioned. Especially as its "nature" is not discussed. But otherwise it is completely in line with the flat stereotypical thinking that is modern physics.

The physics and cosmology
The search for existence foundations came in ancient times take place after two lines. A physical and a mathematical. The ultimate ground, Chaos, the four elements (plus a fifth, then ether) and Democracy atoms and empty space on the one hand and the Pythagoreans' "all is number."

Quantum physics was born with the discovery of energy quantization and constant h. Kirchhoff discovered around 1860 that there is a universal function of only frequency and temperature that can describe their thermal radiation. Planck partially solved the problem in 1900 by introducing a quantum unit h (Planck's constant), which was of great importance for quantum physics development.

But neither he nor anyone else has later managed to define and derive this constant, as little as the light constant c or the gravity constant G. This is because the so-called quantum mechanics are only a set of mathematical rules and tricks, rather than insight and understanding. Quantum physicists even brag about this lack of understanding, they mean that the nature is bizarre and incomprehensible to its very nature. "Everything is riddles" is their credo.

Who can be happy about such? Yes, the columnist in the daily press. They love black holes and the idea that the universe is incomprehensible. Unfortunately, their worldview in some cases become scientific "truths" truths then banked and hammered into all media and then end up in textbooks.

The reality of the physics is like the god of Janus - it has two faces. Here a real and an imaginary (we can imagine!).

Niels Bohr, one of the quantum physics fathers, said, for example, that the "who is not ill-affected by quantum physics has not understood it" and "Nothing exists until it is measured". Something that is usually perceived as some sort of profundity, but that is only a lack of a well thought-out theory of knowledge.

Einstein tried to find the logic of quantum theory many times, but did not work well. Time was not ripe for this. In his view of science, as in Maxwell, concluded that it is necessary to form a picture of what is happening. To understand a steam engine, you must imagine how the steam came into a cylinder to quickly expand and push a piston in front of it. How this drove a flywheel driving other wheels etc. and all this because water was boiled in a large container like ... etc. Therefore, in order to understand the electrical and magnetic phenomena and other natural processes, both Maxwell and Einstein represented mechanical gears, belts, feathers, waterfalls, wheat fields, lifts, carousels, etc.

Something that has since been regarded as both childish and ignorant in the illumination of modern physics with its indescribable bizarre quantum world, where only statistical and mathematical descriptions pertained and function. Analogue with wheat fields, which "bends for the wind", which gave birth to the field concept, is moreover an alternative to the mechanistic perception and refers to a more complete description of reality. Maxwell, however, did not see the fields as pure mechanical things, but more like esoteric irreversible ether particles.

Certainly, some important facts about the nature, space and our universe were missing at the turn of the century and far into the 20th century. But really not today with many satellites full of instruments spinning around the Earth and other planets. So, instead of mystifying the physics and stating that the quantum mechanics and wave mechanics are the last word and the "Great Truth" about our world, one should have resigned, seen time and thought that new facts might solve the problems we have now. Which was Einstein's setting. For him, mathematics was not the starting point but nature, and he sought non-mechanical solutions. Where even Maxwell had made valiant attempts. But as I said, time was not ripe for this, at least in the 19th century.

Metaphysical elements into physics

Enough of that now. The whole of modern physics project was to derail the 1920s. Since one could not quickly get a clear picture of how the light and light photons worked, the physics community abandoned the idea of understanding. With Niels Bohr and Werner Heisenberg in the lead, metaphysical elements were introduced into physics.

> The progress recently made in nuclear physics and elementary particle physics / ... / has still not fundamentally solved the epistemological crisis that quantum breakthrough years 1923-27 gave rise to. /.../ The problem, which consists in unequivocally producing the microphysical reality in the form of a space-time model, remains unresolved, and probably the ambition to solve it must be stated. The reality of the physics is like the god of Janus - it has two faces[8].

This, according to a distinguished French physicist. The more famous physicist and philosopher Karl Popper has written and analyzed how and why modern physics derailed in the 1920s. This is his view on the matter:

> Today, physics is in a crisis. / ... / But there is also another aspect of this crisis: it is also a crisis of understanding. This crisis of our understanding is about as old as the Copenhagen interpretation of quantum mechanics.[9]

Creator of the field concept - Faraday's lines of force - Michael Faraday is considered to be. As we will see, the photons and neutrinos we treat here are a development of the field concept. In my theory, these force fields, are both real and imaginary in a certain way. They also have definite dimensions and are always in certain contexts.

With our new theory and approach, this also applies in principle, even though its imaginary world is a matter for itself. However, it is possible to make pictures of all processes as well as give them mathematical values. And then we do not mean statistical point marks. This line of force field line or power line is, in my opinion, a combination of real time and imaginary distances and dimensions, giving the quantized force = $iF = i\sqrt{h/2\pi}$. Points out, however, for Einstein, they were only real and of a continuous nature. The crisis Popper saw

---

[8] Alfred Kastler. French Nobel Prize in Physics 1966. *This strange matter.*

[9] Karl Popper: *Quantum theory and the schism in physics.*

in the new physics was, in his opinion, essentially due to two things: a) the interference and the penetration of subjectivism in physics; and b) that the idea prevailed that says that quantum theory has reached a full and final truth. This is often the content of any criticism of the Copenhagen interpretation, and it was Einstein.

The natural dialectical particle theory

However, as said, all the needs of quarks and gluons (adhesive particles) disappear with the new particle theory, while many others do not turn out to be fundamental at all but are combinations of photons and neutrins. This applies, for example the electron and proton. First, however, we need to figure out and answer the question Einstein asked long ago. It is a little strange that professional physicists do not seem to have deepened in this extremely important question for the whole of physics. So, once again:

> All these fifty years of conscious brooding have brought me no nearer to the answer to the question, 'What are light quanta?' Nowadays every Tom, Dick and Harry thinks he knows it, but he is mistaken. (Albert Einstein, 1954)

What are then light quanta? The new particle theory is based on a new way of looking at nature, also a new geometric model of how the photons work and, therefore, what a light quantum is. We perceive our world, our being, as entirely mechanical. But this is not always true with experiments and many other observations. The being of light is not mechanical, it is in another being, in a non-mechanical world called electromagnetic. In order to describe this non-being, this electromagnetic and non-mechanical world, complex numbers and the imaginary device are needed.

But I would say that this applies only in the world of "pure" mathematics and religion, where the Absolute, Infinite and perfect God to 100 percent rule. We thus do not forget about the absolute and the relative. We must be taken from pure mathematics and physics. It is religion, where the absolute, infinite and complete god controls 100 percent. This monolithic black and white thinking is the modern physics mortal sin.

The light also has a mechanical side, it has an impulse moment. Neither light nor we, live at 100% of its being. Nor is the non-being to 100% absent. The theory, a natural and dialectical particle theory, answers Einstein's question and gives us new universal building blocks. From this photon, all other particles can be derived in one way or another. But then we have to go to the basics. This causes a number of other problems to be investigated and explained. Far from everyone, but on the limited space, the answer looks like this:

1) First, the question of what light and all other electromagnetic radiation needs to be explained. This is especially about the familiar light photon. Modern physics has not answered this question of light quantum. Neither J C Maxwell nor Max Planck. Not even by Einstein himself, as can be seen from the quotation above. You might think so, as many do, when he even presented a theory about the photoelectric effect in 1905, a theory that was awarded with a Nobel Prize later. Because he asks about this 50 years later as we look at the quotation above, that's obvious. That people in general (every Tom, Dick and Harry) think that everything that concerns the photons of light and the quantization mechanism are completely clear, that's another question. This also applies to many physicists.

So I can proceed to:

2) A new model of the light, a new model and the theory of what a photon of light is. When this is clear, then only then can the question of what light quantum is answered. Indeed, there is a lack of consistency between theory and experimentation with the common model of light photons. Both the so-called double-slit experiment and polarization experiment falsifies obviously the prevailing model. Something must be wrong, seriously wrong. It is in fact a bit strange that, instead of putting the current model and its theory in question, you conclude that there is something wrong with nature! Strange, bizarre and illogical as it is, is the judgment! Bohr, Heisenberg, Born and their thoughtless apologists really sat on high horses!

3) This requires a new model of the light photons – a further outline, one that encompasses our entire world and reality. The theory must be able to describe how each photon in any frequency range works and, after a completed revolution (2π), can be given this general formula:

$$f* \, i^4 \, h * e^{-i2\pi} = f*h$$

The photon thus switches between building up a magnetic field perpendicular to an electric which correspondingly reduces - discharged. In the next step to discharge the magnetic energy which is converted into electricity. The photon therefore functions as an electric/magnetic oscillator, an LC oscillator, where the electric perpendicular to the magnetic and also indicates the polarization direction.

This therefore appears to swing up and down while the sum of the fields is constant. When the magnetic is close to zero, the electrical is close to its maximum value. And vice versa.

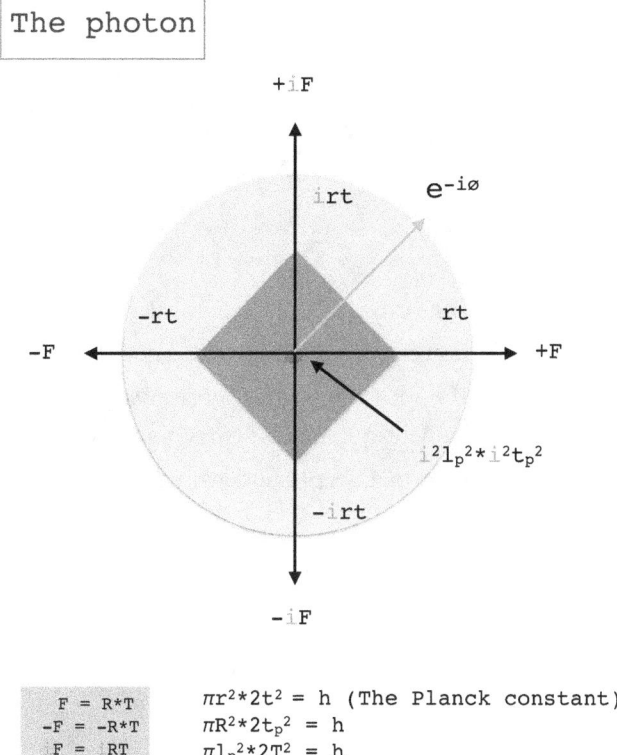

The number of steps that the green vector / arrow can take is: $2\pi / 2me = Z \approx 3.5 * 10^{30}$.

This quantity is always equal to an angular momentum of Planck's constant h.

Two conditions must therefore be met before such a "quantum jump" can be made. First of all, the vector is in the real zone. (See some dimensions and patterns of the photon below!). The second that the product of the circular surface (πr^2) and the square ($2t^2$) equals Planck's constant h. The rectangular surface of the figure on the next page thus corresponds to a square-dimensioned area. The larger this, the smaller the amplitude and the shorter the distance between each photon in a photon train or photon beam, which in the normal physics corresponds to the wavelength (λ).

A light quantum is then divided into a large number of smaller quantities. At a higher cosmic level there is yet another quantity, proportional to Planck mass (m_P). This quantum exists at different levels must be

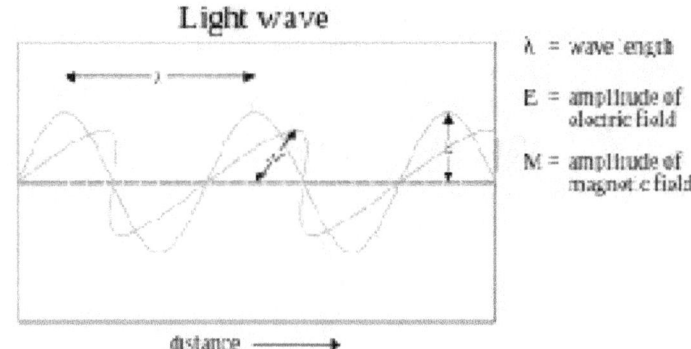

In 1900, Maxwell's theoretical model of light as oscillating electric and magnetic fields seemed complete. However, several observations could not be explained by any wave model of electromagnetic radiation, leading to the idea that light-energy was packaged into quanta described by E=hv. Later experiments showed that these light-quanta also carry momentum and, thus, can be considered particles: the photon concept was born, leading to a deeper understanding of the electric and magnetic fields themselves. (Wikipedia.)

included in the new quantum theory. Thus:

$$Z^4 * i^4 * h = Z^4 * \pi R^2 * 2T^2$$

It also applies to the product of minimum and maximum areas:

$$i^4 h = \pi r^2 * 2t^2 = \pi R^2 * 2tp^2 = \pi l p^2 * 2T^2.$$

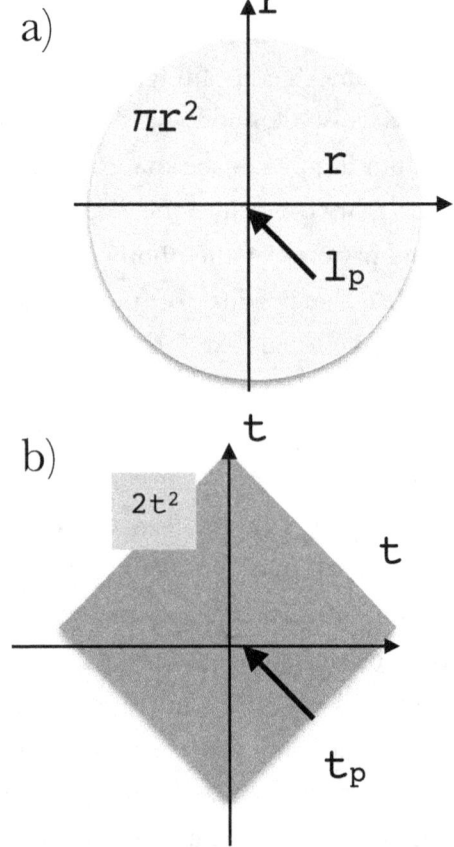

a)

b)

For a particle with the spin h / 2, for example, neutrinos, electrons and neutrons, the relationship:

$$i^2 h * e^{-i2\pi} = h / 2$$

Here met apparently not the condition of a completed lap. So, where does the other half of the angular momentum (h / 2) path? Well, it means that the other half of the neutrino moment now binds together - stick together - the two components I am talking about.
The force + F or -F becomes active!

Clearly, the need for a particular "gluon particle" - the gluon clearly disappeared. Oops!

Photon dimensions and components
The photons thus have two components. Inserted in a two dimensional Cartesian coordinate system (CS) looks like this:
a) a surface with the metric dimension (r^2 square meters) and one with
b) temporal (t^2 square seconds). If we let the two components coincide, we have the graphics shown next page in c).
d) Here the coordinates of the photon becomes *four-dimensional*. And here the photon even becomes *imagin*.

c)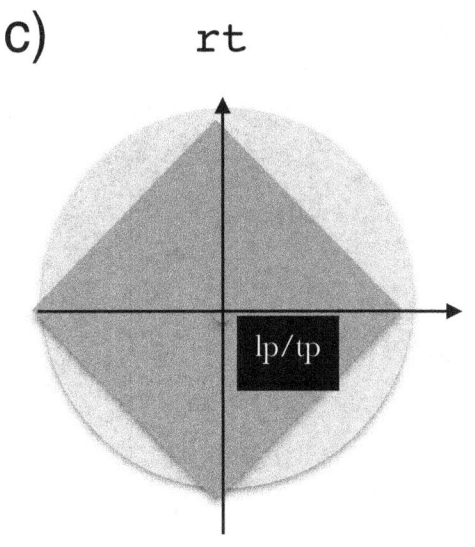

d)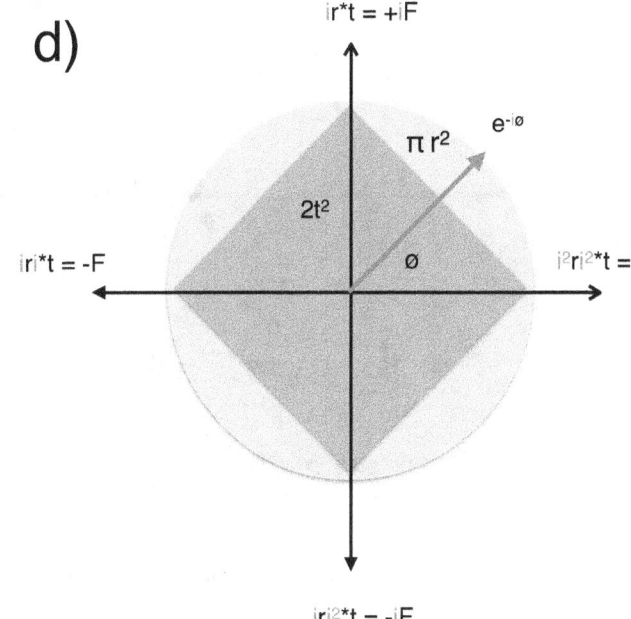

The red arrow in d), right, represents a vector. Right now about 45 ° slope, ie π/4. The whole lap is 2π. At 90 ° it has passed the imaginary zone of the photon, at 180 ° the anti-real and at 270 ° the anti-imaginary zone. When it comes past this zone, the photon is in a real state, but only one passive one. Not until it's very close to 360 °, ie 2π, it achieves the condition of a full turn and has the ability to become completely real, ie, get a mechanical character.

> The red arrow in d) – a vector – shows a photon whose switch between the electrical and magnetic fields moved about 45 degrees of 360. The sum of the electrical and magnetic fields is constant. (See formula).
> It is this constant shift which causes the arrow / vector to circulate like a clockwise pointer – which becomes sinusoidal from the side – *as therefore the common physics believes is a wave*
> **Formula: $\cos^2\phi + \sin^2\phi = 1$**

The condition is that the whole impulse moment *h* is fulfilled. At each measurement, this condition becomes – the mechanical -– fulfilled simply because a measurement, a detection, etc. is a *mechanical* process (not an electromagnetic). If we try to measure somewhere between almost zero and close to 2π, nothing happens. Now note that the photon just described are in a bound state, not in electromagnetic and free state. That the photon is bound means that it is linked to maybe another photon or maybe a neutrino …

The geometry of the photons can be described with a four-dimensional coordinate system with two vertical and two horizontal imaginary axes, which means a complex

> The photon is an electromagnetic LC circuit, an oscillator which in its free state is imaginary and has the shape of a very thin CD, a disk. It is this condition I sign with **i**. (For clarity with red designation).
> Each photon may have some energy, which indicates the ratio between the radius r and t. The frequency (*f*) then becomes *f* = c / r and the energy proportional to this. In the graph above we can see how this can be described. The larger the $2t^2$ (the time area), the smaller becomes πr^2 (the area in meters), which simultaneously means greater frequency and shorter wavelength, ie. r is less the higher the frequency. *The photon therefore, as we see, has an internal structure …*

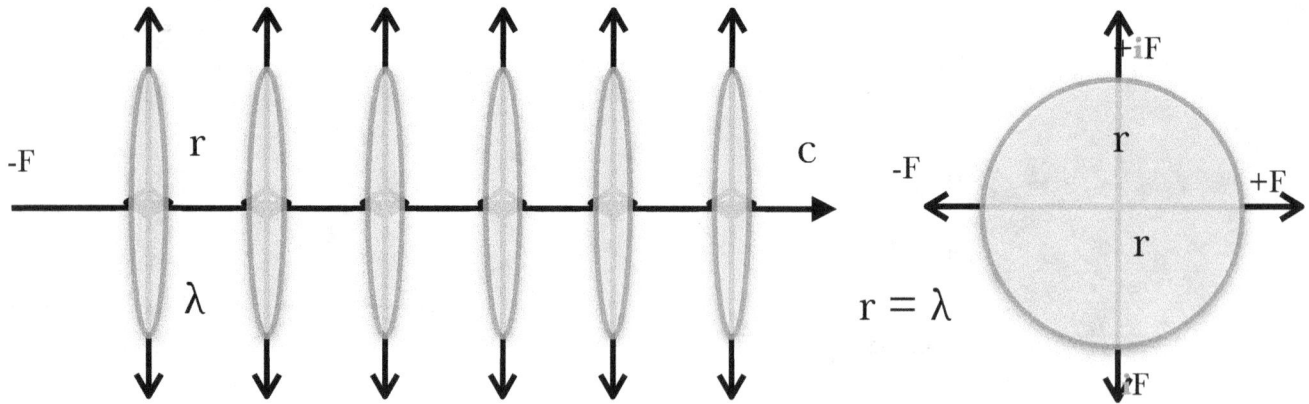

(The new) Light wave where the photons here are angled slightly on the side and the last straight from the front

system. It is a description of the photon in a *free electromagnetic state.*

Such an individual photon is what we usually consider as a light quantum. Each constituent of a light beam, a radio wave, an X-ray, etc., is such a light quantum. In the graphics d) we have more of the photon data and we also see how different forces, $F (= R * T$ or $r * t)$, affect and are affected by the photon. As mentioned earlier, they can be imaginary, iF, anti-imaginary $-iF$ and real $+F$ and negative $-F$.

As previously mentioned, it is not wrong to see the different F: one force with universal extent, as a sort of antennas or feelers. Sensors that detect and affect the environment, constantly maintaining contact with, retrieves and delivers information from and to the outside world. That such ability has an imaginary component makes this possible. In *ordinary physics called field.* They have a "spooky distance effect," as Einstein did not believe in, but as there are actually experimental evidence in fact. Below we see how a light beam or a train of photons — where the graphics now *are simplified and stylized as below* — can be visualized.

The photons are interconnected, thus forming a kind of train, thanks to their forces $\pm F$. They are loosely joined, but though. (For educational reasons, here the blue squares are not written.)

If the front of a photon produces the force F, its back is -F. This causes two or three in a row to form a kind of balance so that their distance becomes equal to its radius r. A distance as the normal physics calls the wavelength. But because the photon is not a wave, of course, this term is irregular.

Thus, in the graphics above, a train of photons is visualized. We may think that we see them at the moment they have potentially completed an entire revolution with an accomplished motion moment *h* which can be detected by a mechanical impulse p. Viewed from the side the photons appear in a row as a fence, with the distance ***r*** — a distance the normal physics calls wavelength and usually denote **λ** (lambda). Thus **λ = r.**

—

After these preparations, we might now able to approach our main object in this paper: the electron. Then we will study how electrons (and positrons) occurs. First, the common approach and theory. Consider this process:

Thus, we see how a gamma-ray hits an obstacle and as a result of this "collision" an electron positron pair is formed. Wikipedia:

> Pair production is the creation of an elementary particle and its antiparticle from a neutral boson. Examples include creating an electron and a positron, a muon and an antimuon, or a proton

and an antiproton. Pair production often refers specifically to a photon creating an electron-positron pair near a nucleus. In order for pair production to occur, the incoming energy of the interaction must be above a threshold in order to create the pair – at least the total rest mass energy of the two particles – and that the situation allows both energy and momentum to be conserved. However, all other conserved quantum numbers (angular momentum, electric charge, lepton number) of the produced particles must sum to zero – thus the created particles shall have opposite values of each other. For instance, if one particle has electric charge of +1 the other must have electric charge of −1, or if one particle has strangeness of +1 then another one must have strangeness of −1.

The probability of pair production in photon-matter interactions increases with photon energy and also increases approximately as the square of atomic number of the nearby atom.

Quantum field theory

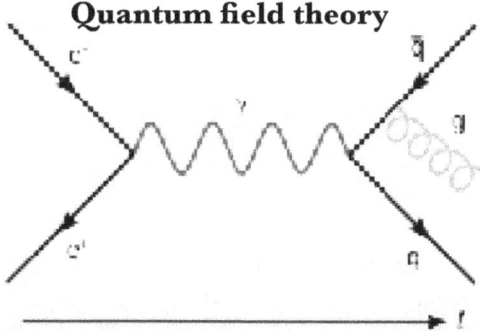

In this Feynman diagram, an electron and a positron annihilate, producing a photon (represented by the blue sine wave) that becomes a quark–antiquark pair, after which the antiquark radiates a gluon (represented by

Photon to electron and positron

For photons with high photon energy (MeV scale and higher), pair production is the dominant mode of photon interaction with matter. These interactions were first observed in Patrick Blackett's counter-controlled cloud chamber, leading to the 1948 Nobel Prize in Physics.

If the photon is near an atomic nucleus, the energy of a photon can be converted into an

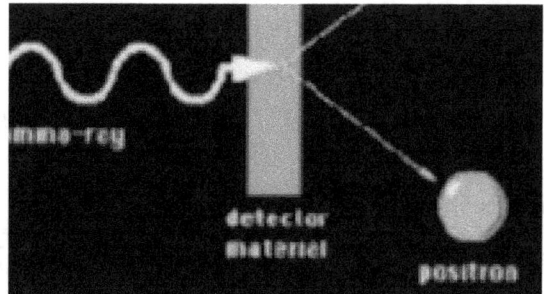

electron-positron pair:

$$\gamma \to e^- + e^+$$

The photon's energy is converted to particle's mass in accordance with Einstein's equation, $E=mc^2$; where E is energy, m is mass and c is the speed of light. The photon must have higher energy than the sum of the rest mass energies of an electron and positron (2 Now here is the result of the photo beam "collision" with an obstacle (an atom); three photons layers on top of each other. Then quickly decay into an electron and a positron.× 0.511 MeV = 1.022 MeV) for the production to occur. The photon must be near a nucleus in order to satisfy conservation of momentum, as an electron-positron pair producing in free space cannot both satisfy conservation of energy and momentum. Because of this, when pair production occurs, the atomic nucleus receives some recoil. The reverse of this process is electron positron annihilation.

The formula below then gives 45⁰ that both cosine in square and sinus in square will be 0.5. The sum will then be 1.0. For 90⁰, cosine becomes 0.0 and sinus 1.0. This means that the electric field is maximum — 100%. By 135⁰, sinus and cosine again become 50%. 180 degrees, the electric field becomes 0% and the magnetic maximum is all — 100%. Etc. The vertical electric field thus switches between a minimum (0%) and maximum (100%) over time, which makes it become sinusoidal. Hence the illusion that the light is like that — wave formed.

$$\cos^2\emptyset + \sin^2\emptyset = 1$$

The article in Wikipedia also contains texts on kinematics and energy transfer, but we leave to the reader that if interested to study further. Clearly, the mechanism that can explain what happens when a sufficiently powerful electromagnetic gamma undulated ray is converted into, or creates, two opposite forms of electrons is missing. We do not get a picture of what really happens. It is as if the "obstacles" mentioned (a detector material, a nucleus) serves as a screen concealing a secret mechanism, a kind of magic art.

Another similar idea is given by the next picture next page, but uses the term materialization of the gamma ray photons.

From the picture the physicists provide of this process, it is clear that the electron is

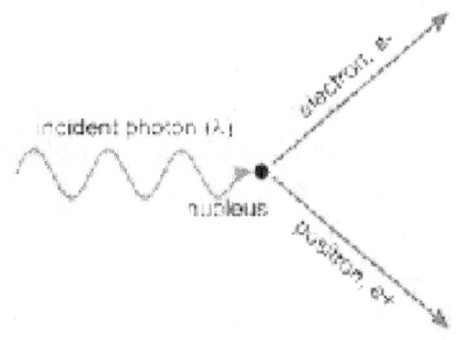

assumed to be a small electrically charged ball. They also have no other view on electromagnetic radiation than it is in the form of a wave. In the physics we have today you only see the forest – not the trees. Even less its branches and roots. Or you are like a pediatrician who does not know how babies are made (materialize). It does not get any better with Feynmans graphics from Wikipedia, rather it helps to enhance the tricky device, and haze in modern physics. See the example below where an electron and a positron annihilate, *producing* a photon. And mixing quarks in the whole! Huja!

I will now show a completely different picture of the process, an image that not only shows an external occurrence, but one that both shows and *explain* the process in its interior from beginning to end. The reader can then choose what he/she prefers, and do provide a greater understanding of the process.

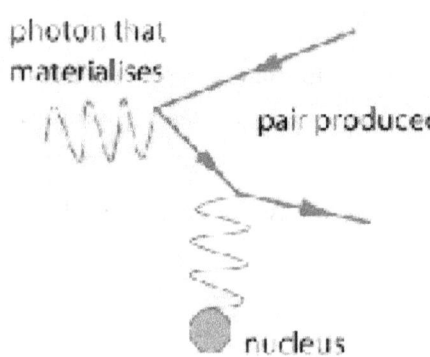

Another exempel of how the common pfysics lock upon these mechanisms

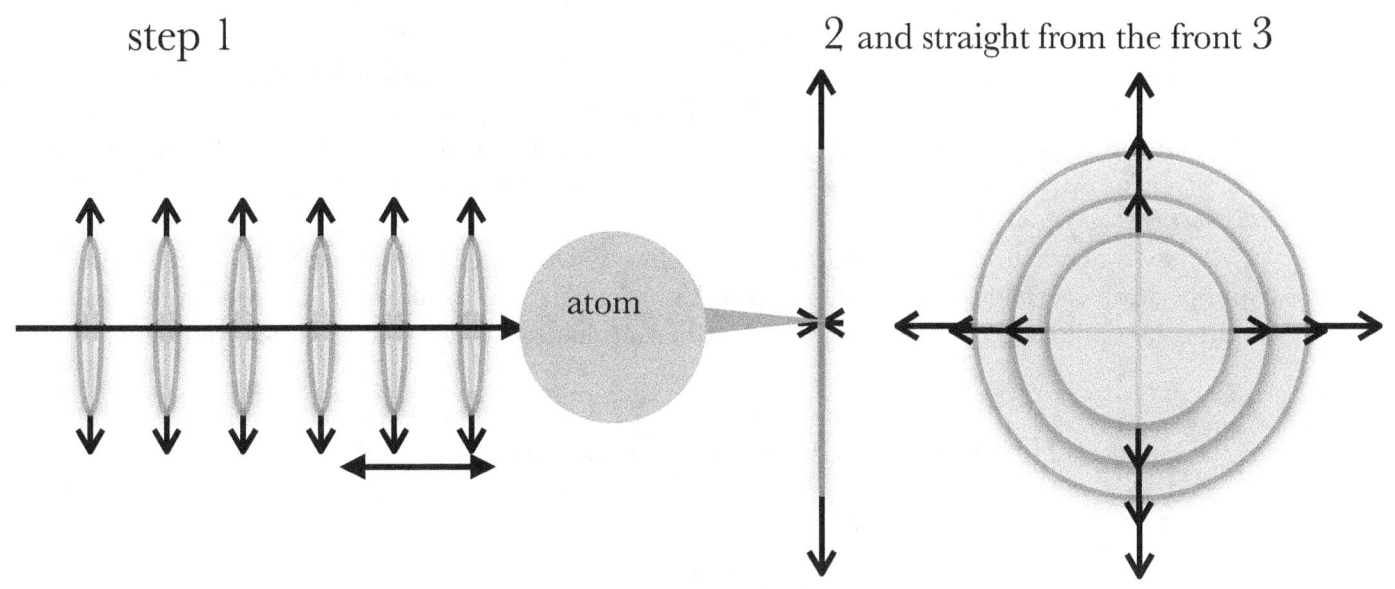

For example, three of these photons are superimposed. At the same time, they change aggregation states from the imaginary to the mechanical.

After a very short moment as a kind of a superphoton, it breaks down into an electron positron pair

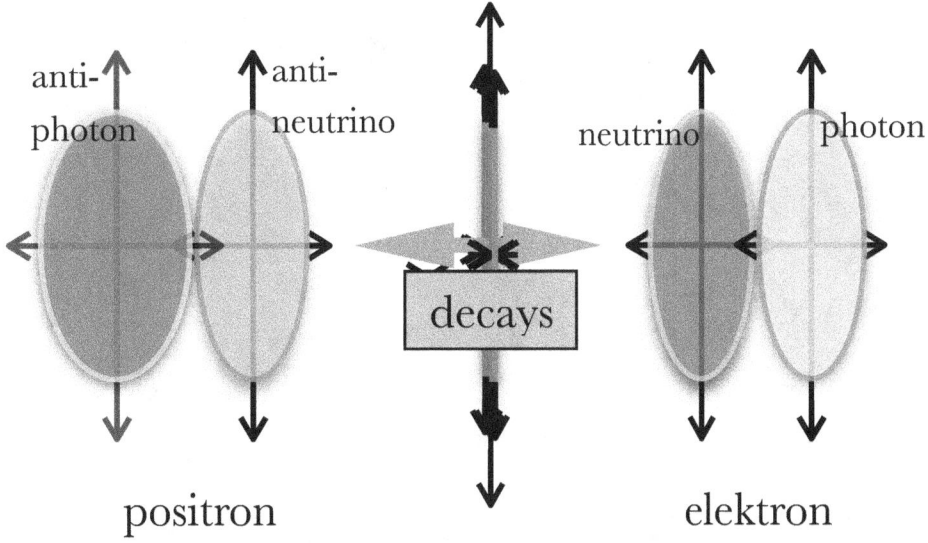

Now here is the result of the photon beam "collision" with an obstacle (an atom); three photons layers on top of each other. Then quickly decay into an electron (right) and a positron.

But here we now see above the internal structure of the electron, and it is anything but a point, namely that the electron is a compound object, a combination of a neutrino and a photon, while the anti-electron or positron as it is also called a composition of an antineutrino and an anti-photon. That it actually is so I will soon demonstrate and prove at a further study of a number of very well known physical reactions. This is how electrons can be produced.

What about the protons?

Below is another (different) picture of the three or four steps like the formation of electrons and positrons, a truly fundamental process. Visualization (which allows us to understand) can of course be done in many different ways, but the mechanism is always the same. We now have a

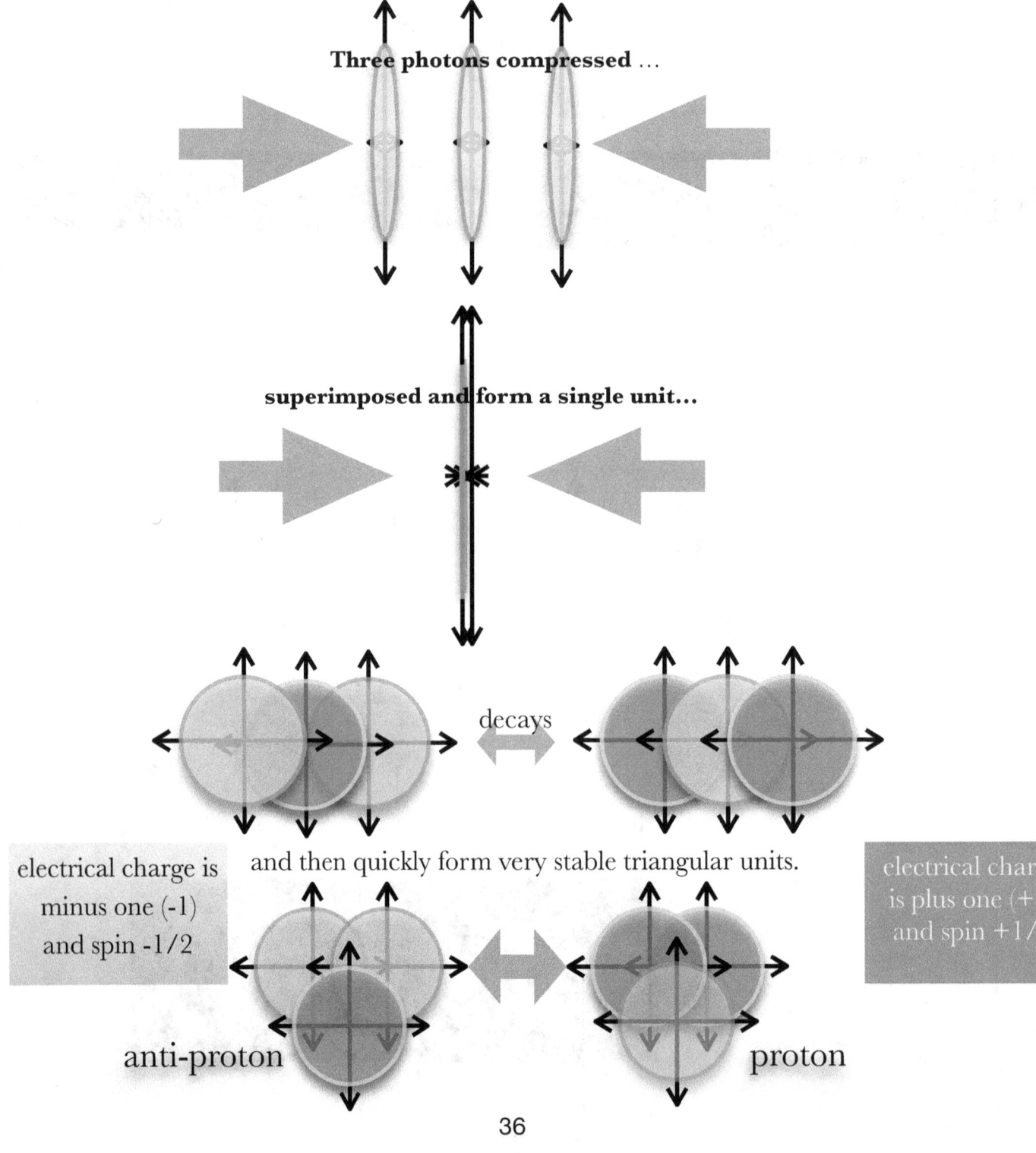

Now, how does this mechanism work when it comes to neutrons? Yes, we have just seen on the previous page. As we can see now 7 photons are required. We can see it as a proton receive an injection of a photon and anti-photon. This makes four photons. If we add the three required for only the protons, then the sum is understood to be seven (7).

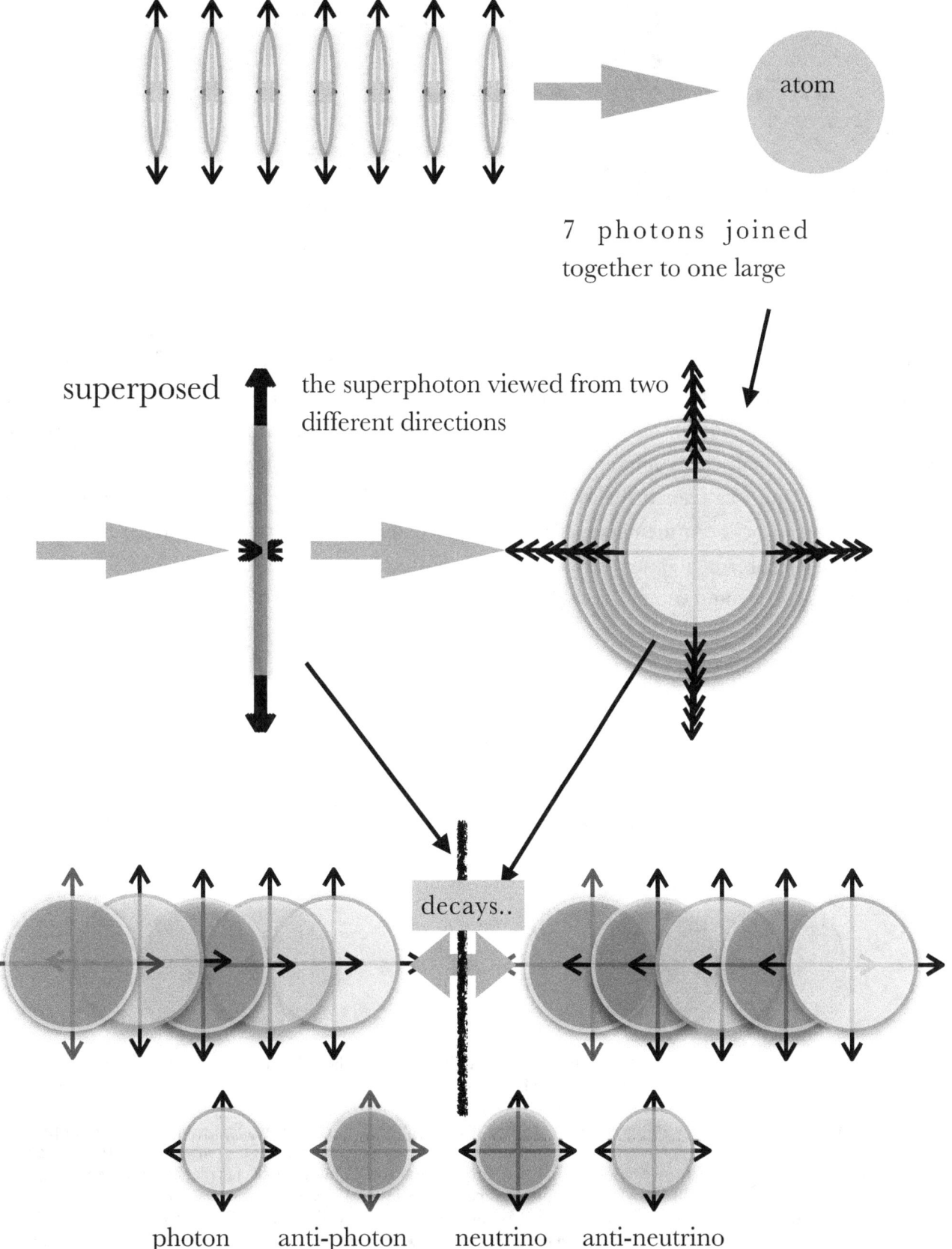

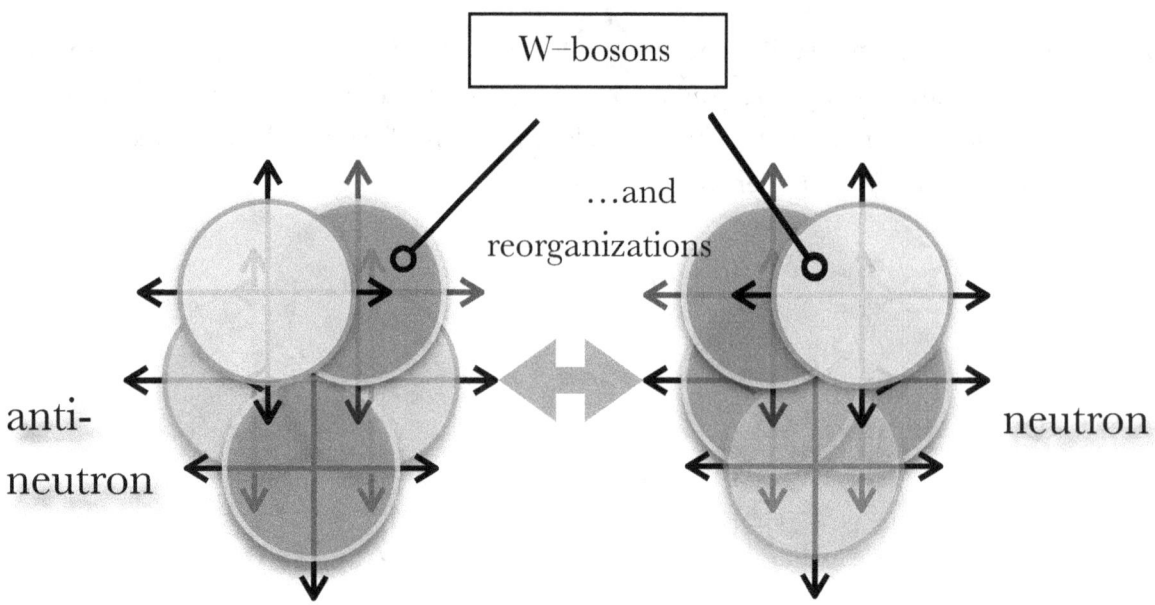

What are we seeing? If we start with the neutron, it is composed of a proton ( ) and a *new particle*, the W–boson, ( ) which consists of a pair of photons (or carefully calculated a photon and an anti-photon). The anti-neutron in turn is composed of an anti-proton ( ) and *the new particle*. This new particle is interesting. It was discovered in the 1980s in connection with the neutron decay and was named the W-boson[1]. We can see it in the chart below. It can have the electric charge +1 or -1. At the neutron is the charge of the W-boson -1, which means that it takes out the positive charge of the proton +1 and causes the entire particle – the neutron – to become neutral. The structure and composition of the neutron used to be studied in its decay as below:

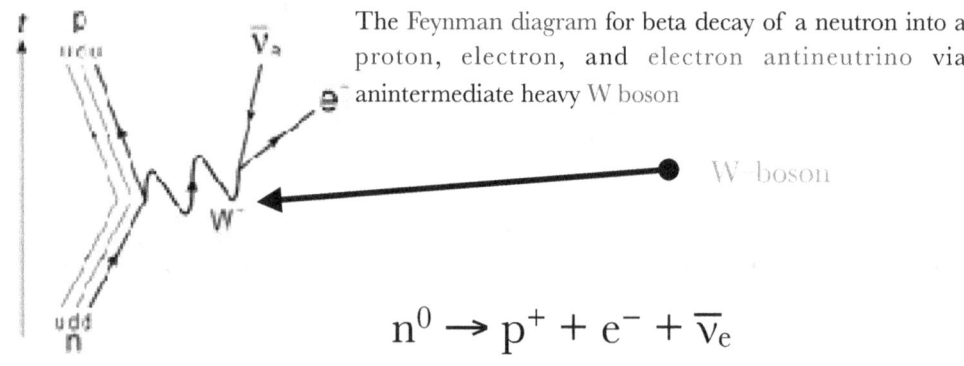

The Feynman diagram for beta decay of a neutron into a proton, electron, and electron antineutrino via an intermediate heavy W boson

$$n^0 \rightarrow p^+ + e^- + \overline{\nu}_e$$

---

[1] Discovered in 1983, the W boson is a fundamental particle. Together with the Z boson, it is responsible for the weak force, one of four fundamental forces that govern the behaviour of matter in our universe. Particles of matter interact by exchanging these bosons, but only over short distances.
The W boson, which is electrically charged, changes the very make up of particles. It switches protons into neutrons, and vice versa, through the weak force, triggering nuclear fusion and letting stars burn. This burning also creates heavier elements and, when a star dies, those elements are tossed into space as the building blocks for planets and even people. The weak force was combined with the electromagnetic force in theories of a unified electroweak force in the 1960s, in an effort to make the basic physics mathematically consistent. But the theory called for the force-carrying particles to be massless, even though scientists knew the theoretical W boson had to be heavy to account for its short range. Theorists accounted for the mass of the W by introducing another unseen mechanism. This became known as the Higgs mechanism, which calls for the existence of a Higgs boson. (This content is archived on the CERN Document Server).

Thus, the process takes place in a single step according to this theory. But this was changed in the 1980s when W-bosonen was discovered. See the graphics which also malfunctioned to it all with quark theory. Also with time arrows crisscrossing.

But first, we will see how other ways normal physics treats the neutron decay. This structure and composition of the neutron can even be studied in its first theory of decay here. Without quark theory. We do a short visit once again to Wikipedia as we have seen on pages 6-7:

$$n^0 \rightarrow p^+ + e^- + \overline{\nu}_e$$

(where p^+, e^-, and ν_e denote the proton, electron and electron antineutrino, respectively.)

Let's study the process in two steps and insert the new W -particle into these.

$$\text{step I}: n^0 \rightarrow p^+ + w^-$$
$$\text{step II}: w^- \rightarrow e^- + \overline{\nu}_e$$

In step I decomposes the nutral neutron into a positively charged proton and the W− particle, which is negatively charged. Together, the charges will be zero. As we can see, the proton and the W-particle live side by side for a short moment. Just what the researchers discovered at CERN in the late 1900s. But the discovery also meant that the W- boson is unstable and decays immediately into an electron and an anti-neutrino in next step, see step II.

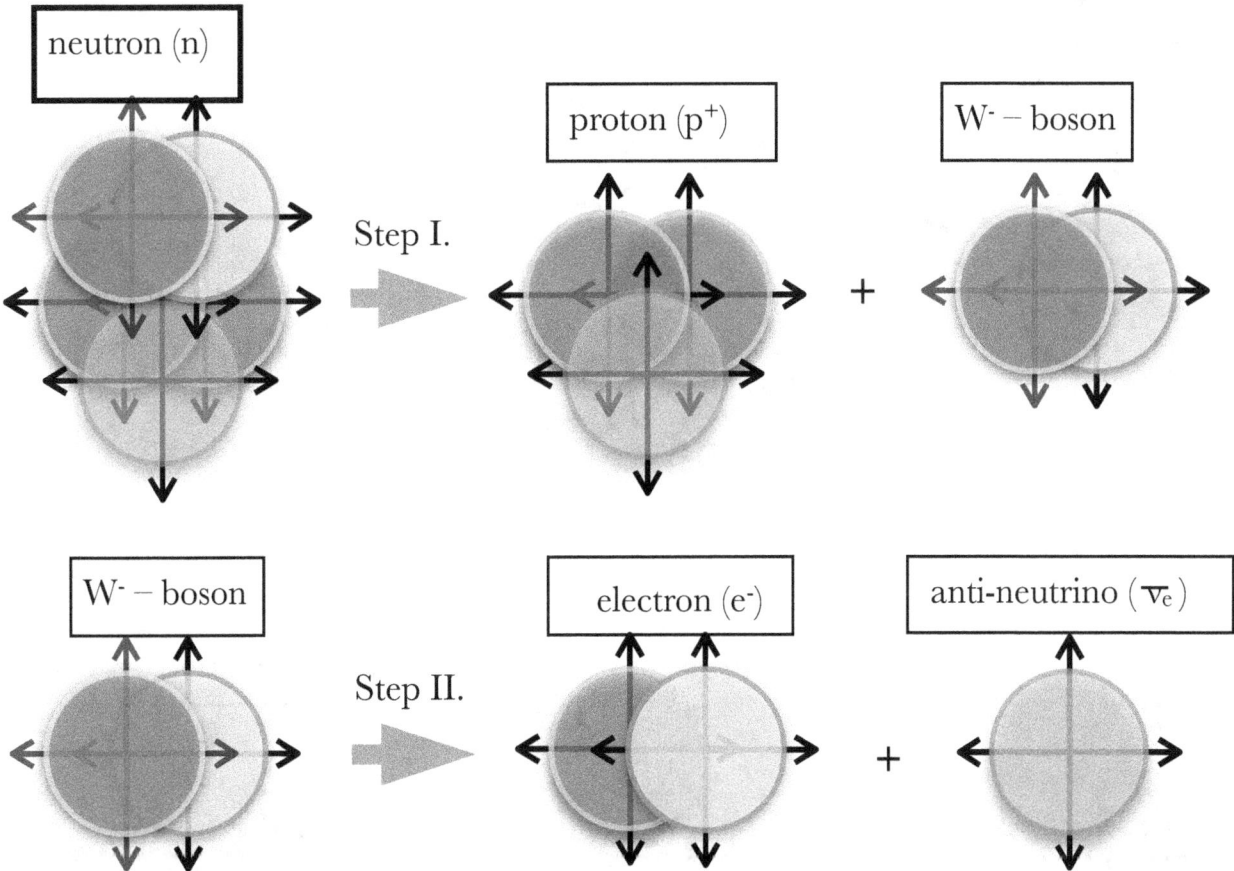

As we see, the W-boson is the key that can unlock the issue not only about whether the electron has an internal structure, but also provides this structure in an exact and convincing manner. Convincing because there are so unique experimental evidence for this.

The W-boson is thus **composed** of an *anti-photon and a photon*. It is its internal structure and intercommunion, its essence, soul and character. An electron defined as a point particle without internal structure and interactions lacks therefore essence, soul and character.

For the $W^\pm$–boson it means that this causes the charge -1 or +1 and the spin 0, because one photon has spin +1, the other must have the spin -1. That the boson gets the charge ±1 is an emergence effect, as previously mentioned. There is the same effect as when an atomic sodium (Na) may combine with one atom of chlorine (Cl). Individually, they are downright dangerous and toxic to ingest (Na corrodes you, Cl destroys your lungs), but together, united in one chemical entity, sodium chloride, NaCl, forming a salt which is good to put in the porridge or have the eggs. Emergens is then when two different things have each of their separate characteristics, but together they get a completely different feature.

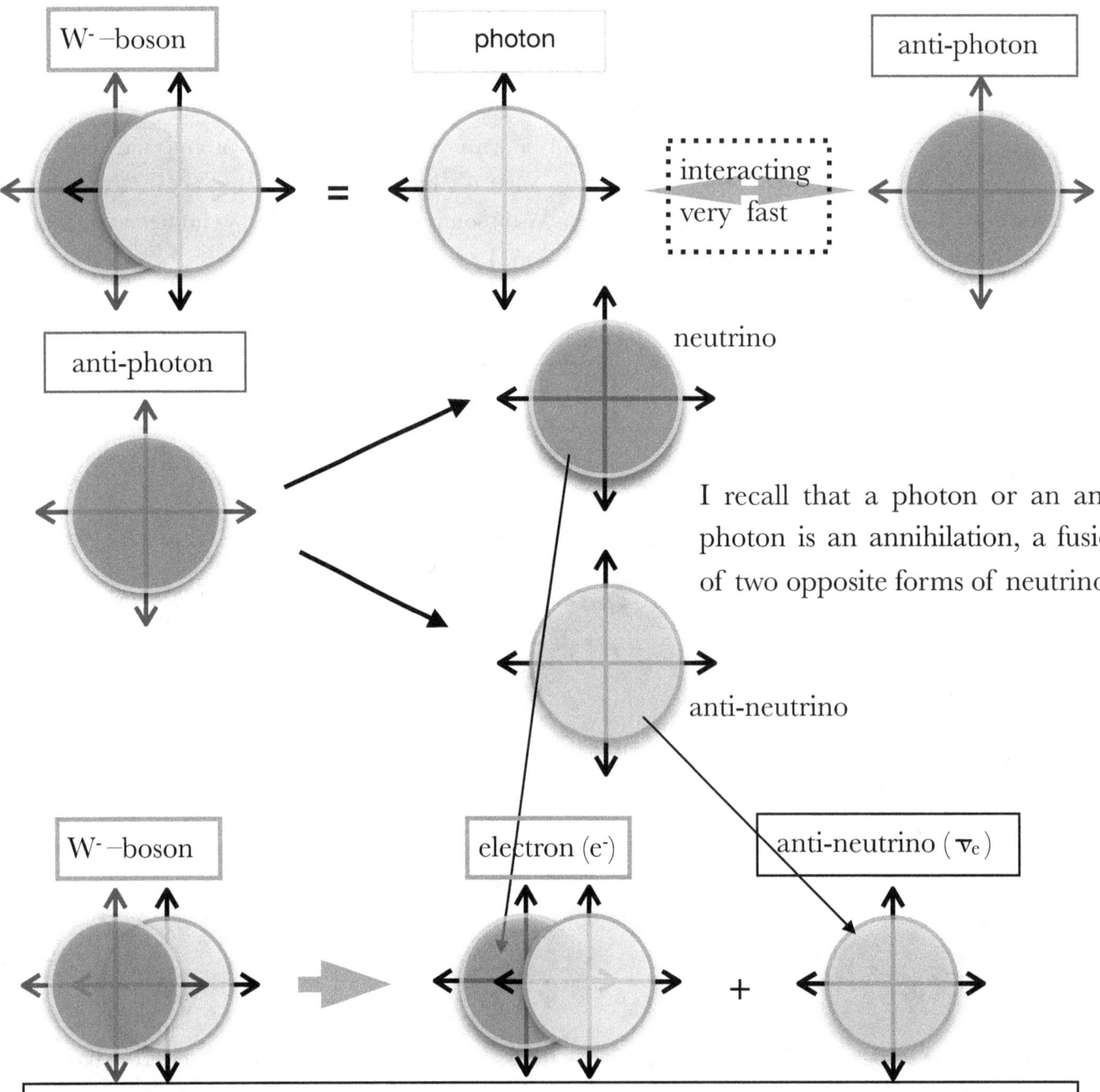

Here a negatively charged boson with the spin zero (0) decomposes to a negatively charged electron with the spin +1/2 and an anti-neutrino with charge zero and the spin -1/2. Everything's in order!

/ ... / Emergent structures are patterns not created by an individual event. Nothing commands a system to shape a pattern, rather, the interaction of several parts creates a complex chain that leads to a certain order. It can be argued that emergent structures are more than its combined parts, because an emergent order is not solely due to the mere existence of the parts – the interaction of the parts is central. Emergent structures can exist in many naturally occurring phenomena, in different subjects such as physics and biology. For example, the form of weather phenomena like hurricanes is an emergent structure. (Wikipedia).

This effect when one plus one does not become two, but something quite different when when quantite becomes quality, explains that they get another new feature, namely mass. This is the secret of the mass. The fact that the electron composition of my theory are known wave particles (if one may say so) as the photon and neutrino also tells us a lot about the electron characteristics of both wave and particle, which is the prevailing view can not at all, in which the electron is a point (or a cloud!).

Well, why does the electron have the electrical charge q it has: $1{,}602 \times 10^{-19}$ As? I'll try to explain that. Let's see what we have for data about the electron itself. It has the mass $9{,}109 \times 10^{-31}$ kg and its frequency f we can get through its energy $E = m_e \times c^2 = f \times h$.

$$f = m_e \times c^2/h = 1{,}28 \times 10^{20} \text{ Hz}.$$

What does that f mean? Wikipedia:

> For cyclical processes, such as rotation, oscillations, or waves, frequency is defined as a number of cycles per unit time. In physics and engineering disciplines, such as optics, acoustics, and radio, frequency is usually denoted by a Latin letter f or by the Greek letter or ν (nu) (see e.g. Planck's formula). The relation between the frequency and the period of a repeating event or oscillation is given by
>
> $$f = 1/T.$$
>
> The SI unit of frequency is the hertz (Hz), named after the German physicist Heinrich Hertz; one hertz means that an event repeats once per second. A previous name for this unit was cycles per second (cps). The SI unit for period is the second. (Wikipedia)

From this we get that $T_e = 1/f = 7{,}80 \times 10^{-21}$ s. The frequency is fluctuations or oscillations of the electron components, i.e. of the photon and the neutrino. They oscillate with time $T_e = 7{,}80 \times 10^{-21}$ s. See the graphics! At time Te, the neutrinon (red) and photon (yellow) change location.

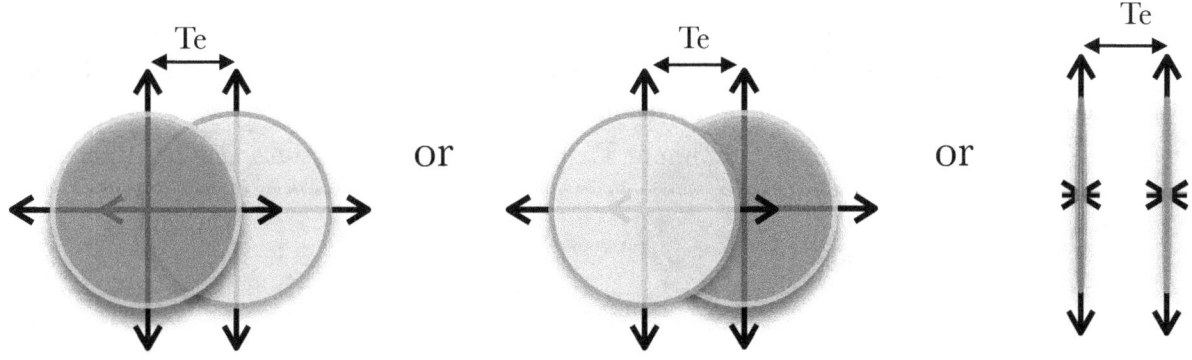

In the case of a light photon with a certain frequency, it means that at time Te, a light quantum passes a certain point. If the frequency is 1 Hz then one light photon passes – one light quantum – per second. At 2 Hz two per second. Etc. In the current case of the electron, its *components* change place within Te seconds. At time 1/Te, they change location 1.28×10^{20} times. Such reasoning is of course difficult to pursue if the electron were a point particle or a cloud!

That T_e is yet another time dimension. We consider once more the photon. Look at page xx8

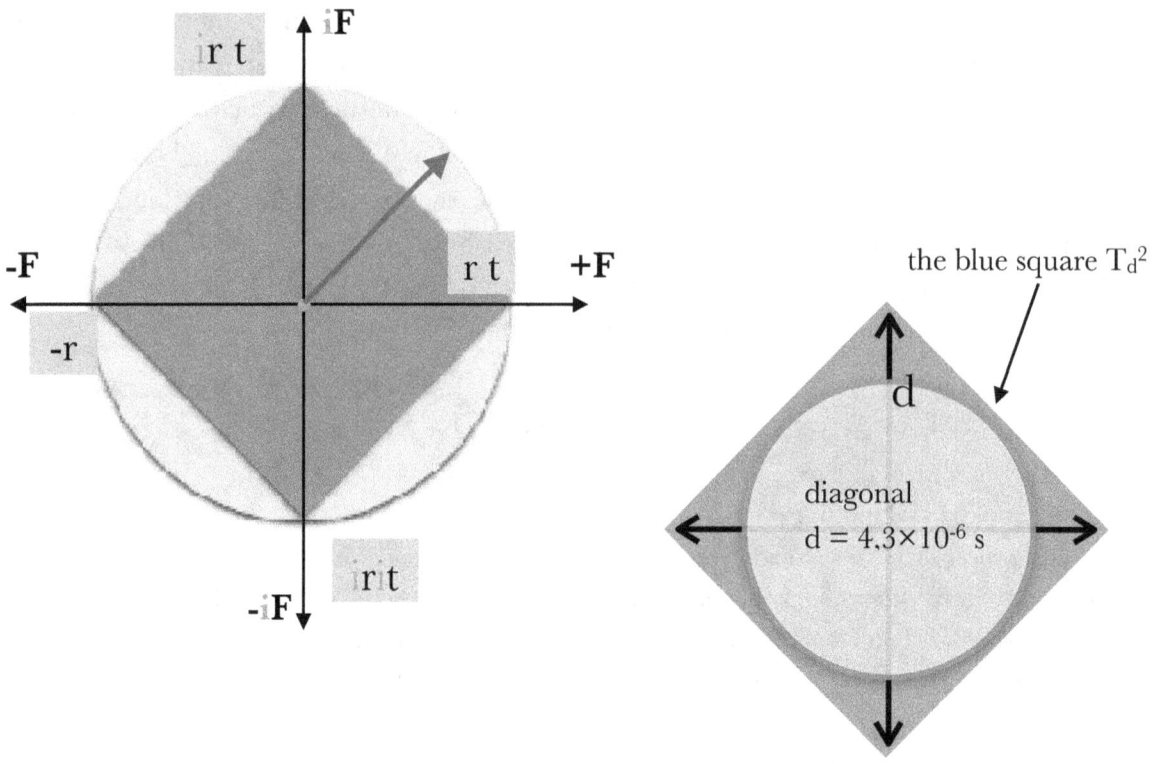

Another important difference is that the original system has no zero point (the red dot), it is a small four-dimensional surface, with the dimensions lp (= Planck-length), and tp, (= Planck-time). R is thus the maximum length and T_m maximum time. Mathematically, it means that:

$$\pi R \times l_p \times 2T_m \times t_p = \pi r^2 \times 2t^2 = h$$

Little h stands for the Planck constant. Welcome to the description of a new world – *the real world!* Unlike the Cartesian artifact as the ancient ruined, completely mechanical "world" that the common physics deals with. (A world which of course constantly held as enigmatic, filled with black holes, quarks , etc., it therefore appear to be. The world I discuss, analyze and compose it all hangs together. As we will see a little further in the text. It sounds like a floskel, but of course it is natural. In my world).

This means that even the product of maximum T_m and Planck time t_p is a *constant*, here called t_{mp}^2.

$$T_m \times t_p = 6{,}32 \times 10^{17} \times 5{,}32 \times 10^{-44} = 3{,}36 \times 10^{-26} = t_{mp}^2 \ s^2$$

Some more important data.

There is even a connection between the electron charge q and the time rotation area $= \pi \times T_e^2 = 2{,}07 \times 10^{-40}$ s². Then time rotational area ($\pi \times T_e^2$) times the angular frequency ($7{,}75 \times 10^{20}$) $= \pi \times T_e^2 \times 2\pi f = 1{,}602 \times 10^{-19} = 1/(\pi^2 \times T_m) = 1{,}602 \times 10^{-19}$ As

The mass of the electron $m_e = 9{,}109 \times 10^{-31}$ kg.
The Planck mass $m_p = 2.176470(51) \times 10^{-8}$ kg

The **two blue areas** - spaces in seconds counted - thus interact and rotate with each other and therefore describe a circle relative to another. The circles surfaces will thus be two circles of radius r = 4.03 × 10⁻⁶ s. (The squares' diagonals). Thus, the two "circles" surfaces become $\pi \times T_e^2 \times 2\pi f = 1/(\pi^2 \times T_m) = 1{,}602 \times 10^{-19}$ As. En dimensionsa nlys ger att eftersom $q = i \times t = 2\pi f \times \pi \times T_e^2$. The angular frequency times Te is thus equal to i, ie ampere strength A. It is thus without dimension.

If the distance calculated in seconds between the electron component is T_e as the ratio *d* between the two equal to $d = t^2/T_e = 4{,}3 \times 10^{-6}$ s. This, then the diagonal (d) of the (blue) square $2T_d^2$.
We have the maximum T_m, the blue diagonal T_d that is part of Planck's constant and so we have T_e, which is connected to the frequency of the electron. Yes, as a yet another kind of time, namely the aforementioned Planck time: t_p. *Four different values of time.* And now we look at this:

$$T_e \times T_d = 7{,}80 \times 10^{-21} \times 4{,}3 \times 10^{-6} = \text{a constant} = 3{,}36 \times 10^{-26} = t_{mp}^2.$$

The electron charge q can we get by realizing that it has to do with its angular frequency. And the maximum time T_m. (Something that a "point" may not like).
Electron mass, we get through it will be a product of its components distance in seconds, times its angular frequency. The product of their time area t_{mp}^2 is in itself is a constant (and of course its square t_{mp}^2 too), so it is only the angular frequency that becomes the determining factor.

To summarize:

Planck constant $= \pi R_m \times l_p \times 2T_m \times t_p = \pi r_{mp}^2 \times 2t_{mp}^2 = h$
Planck (minimum) time $t_p = 5{,}32 \times 10^{-44}$ s
Maximum time $T_m = 6{,}32 \times 10^{17}$ s
The product of T_m and $t_p = 3{,}36 \times 10^{-26} = t_{mp}^2$ s²
Electron frequency $f_e = m_e \times c^2/h = 1{,}28 \times 10^{20}$ Hz.
Electron time $T_e = 1/f_e = 7{,}80 \times 10^{-21}$ s.

Electron angular frequency $\omega = 2\pi f = 8{,}054 \times 10^{20}$ Hz
The electron charge $q = 1/(\pi^2 \times T_m) = 1{,}6019 \times 10^{-19}$ As
Electron mass $= \omega \times (T_e \times T_d)^2 = \omega \times t_{mp}^4 = 9{,}109 \times 10^{-31}$ kg

Plack mass $m_p = t_p \times T_m^2 = t_{mp}^2 \times T_m = T_e \times T_d \times T_m$ kg
$T_d = 4{,}31 \times 10^{-6}$ s².
$T_e \times T_d$ = the product of T_m and $t_p = 3{,}36 \times 10^{-26}$ s².

Is not that wonderful? The product of the shortest and longest time in square is one of Planck's constants? In Planck's constant R_m and l_p above, therefore, the equation corresponds to the maximum and minimum time, now it is therefore about maximum R_m and minimum l_p stretches in meters. Planck length l_p is 1.61×10^{-35} m. The product of this length and the maximum R_m is consequently $r_{mp}^2 = 3.05 \times 10^{-9}$ m. The product of these surfaces = $\pi r_{mp}^2 \times 2 t_{mp}^2$ is thus the meaning of Planck's constant = h.

The maximum and minimum values of length and time show and confirm that everything is connected. Right? R_m and T_m are of course about the world as a whole, but T_m is found as we see in the electron charge q.

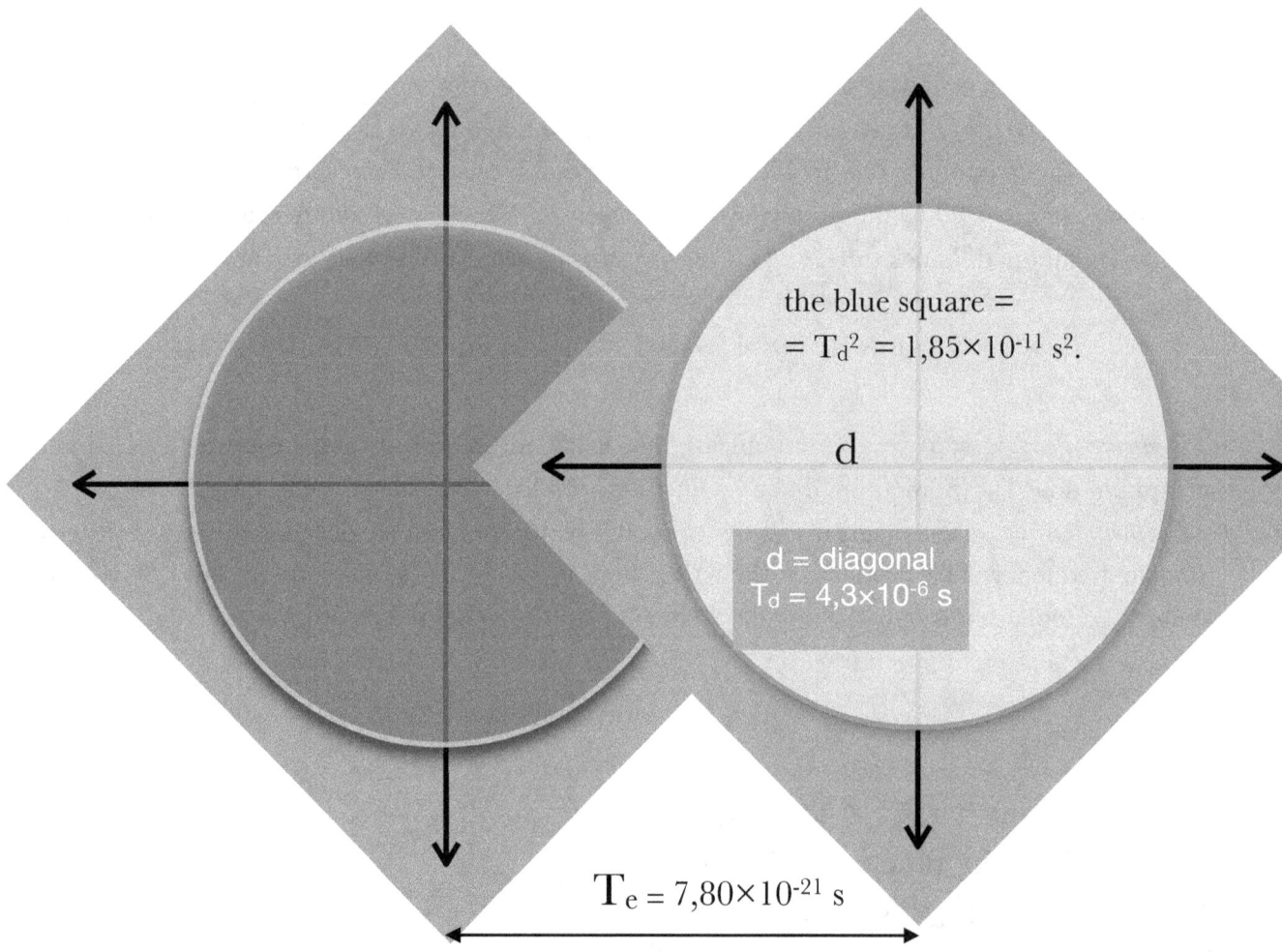

Note the Planck mass (m_p), as we see in $t_p \times T_m$. Etc. The product of the time factor between the two particles of the electron, the distance in seconds between their centers and their time surfaces is a constant = Te × T_d = the product of T_m and $t_p = 3.363 \times 10^{-26}$ s².

What causes the electron mass?

The mass is thus a property that requires a cooperation between at least two components, as we have seen. For the electron, there is an interaction between a photon and a neutrino. The property, or the effect we know as mass, thus occurs when two surfaces with the dimension of time in square interact with each other. The size of this mass, in kilograms, is determined by what distance in seconds counted, or in other words, *with the frequency this change occurs.*

The mass of the electron will therefore be $= T_e^2 \times T_d^2 \times 2\pi f = 9{,}109 \times 10^{-31}$ kg.

($T_e^2 \times T_d^2$ is a constant). Individual particles or components, such as photons and neutrinos, may thus never have this mass property. The interaction and thus the emergence require two or more components. The emergence logically can therefore not occur with a single photon or neutrino.

In experiments one might imagine that such an emergence occurs temporarily in some rare occasional contact between two or more particles. But to assert that it is the mass of, for example, a neutrino you measured, is a premature conclusion of an experimental situation.

The fact that the electron actually has a structure has great consequences for the natural sciences. Next to be investigated now, apart from the proton and neutron, is the atomic model. Now the old planetary atomic model is loose! But I think Niels Bohr would have appreciated it and said: At last! But to this we will return to the next scripture.

Before we leave this we understand that with similar reasoning and with the help of our new constants and their connection with other variables, we can easily understand how the mass of the proton and neutron can be calculated. But now to the fact that the electron has, as they say, a fat cousin called muon. We will now examine how the muon is designed and composed. Wikipedia:

> **The muon** has some similarities with the electron and has therefore been called a fat relative of him, but it is not stable but decays after a certain time. See the decomposition reaction according to the standard version, then consider how it works according to the new approach (in two steps). The muon thus decomposes in a positron and two opposite forms of neutrinos, according to common viewpoints. First few words about this from Wikipedia: The muon then decomposes in an electron and two opposite forms of neutrinos, according to common view. First few words about this from Wikipedia:

> **Muons were discovered** by Carl D. Anderson and Seth Neddermeyer at Caltech in 1936, while studying cosmic radiation. Anderson noticed particles that curved differently from electrons and other known particles when passed through a magnetic field. They were negatively charged but curved less sharply than electrons, but more sharply than protons, for particles of the same velocity. It was assumed that the magnitude of their negative electric charge was equal to that of the electron, and so to account for the difference in curvature, it was supposed that their mass was greater than an electron but smaller than a

proton. Thus Anderson initially called the new particle a mesotron, adopting the prefix meso- from the Greek word for "mid-". The existence of the muon was confirmed in 1937 by J. C. Street and E. C. Stevenson'scloud chamber experiment.

Muons are unstable elementary particles and are heavier than electrons and neutrinos but lighter than all other matter particles. They decay via the weak interaction. Because leptonic family numbers are conserved in the absence of an extremely unlikely immediateneutrino oscillation, one of the product neutrinos of muon decay must be a muon-type neutrino and the other an electron-type antineutrino (antimuon decay produces the corresponding antiparticles, as detailed below). Because charge must be conserved, one of the products of muon decay is always an electron of the same charge as the muon (a positron if it is a positive muon). Thus all muons decay to at least an electron, and two neutrinos. Sometimes, besides these necessary products, additional other particles that have no net charge and spin of zero (e.g., a pair of photons, or an electron-positron pair), are produced.

The dominant muon decay mode (sometimes called the Michel decay after Louis Michel) is the simplest possible: the muon decays to an electron, an electron antineutrino, and a muon neutrino. Antimuons, in mirror fashion, most often decay to the corresponding antiparticles: a positron, an electron neutrino, and a muon antineutrino. In formulaic terms, these two decays are:

$$\mu^- \rightarrow e^- + \nu_e + \nu_\mu$$
$$\mu^+ \rightarrow e^+ + \nu_e + \nu_\mu$$

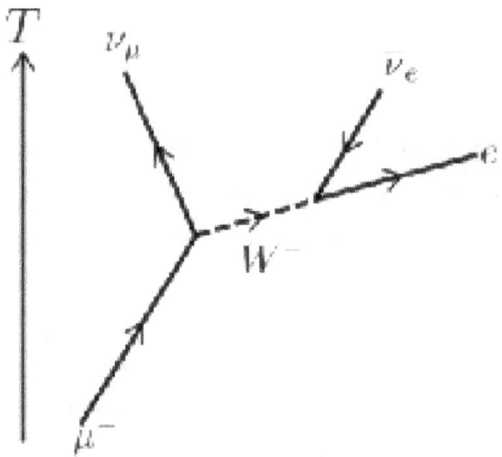

The most common decay of the muon

The mean lifetime, $\tau = \hbar/\Gamma$, of the (positive) muon is $(2.196\,9811 \pm 0.000\,0022)$ μs.[2] The equality of the muon and antimuon 4 lifetimes has been established to better than one part in 10.

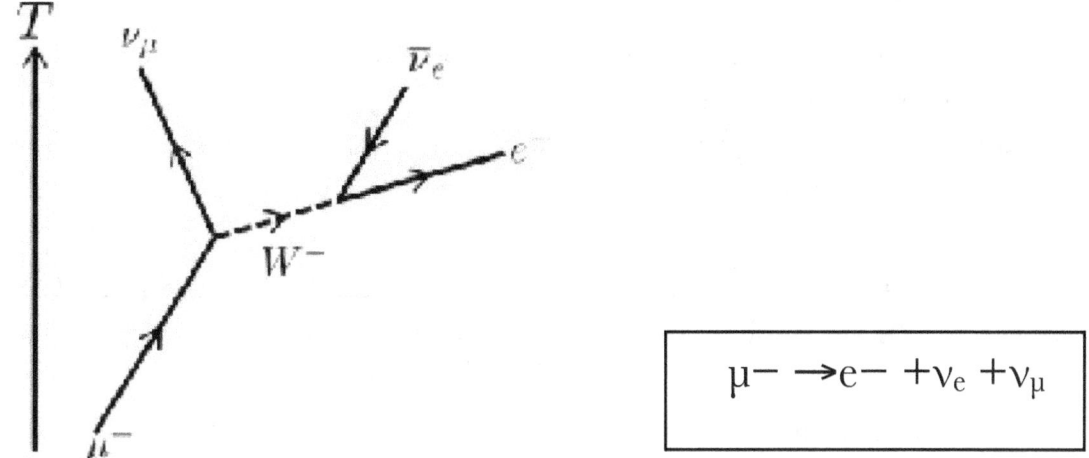

Step 1

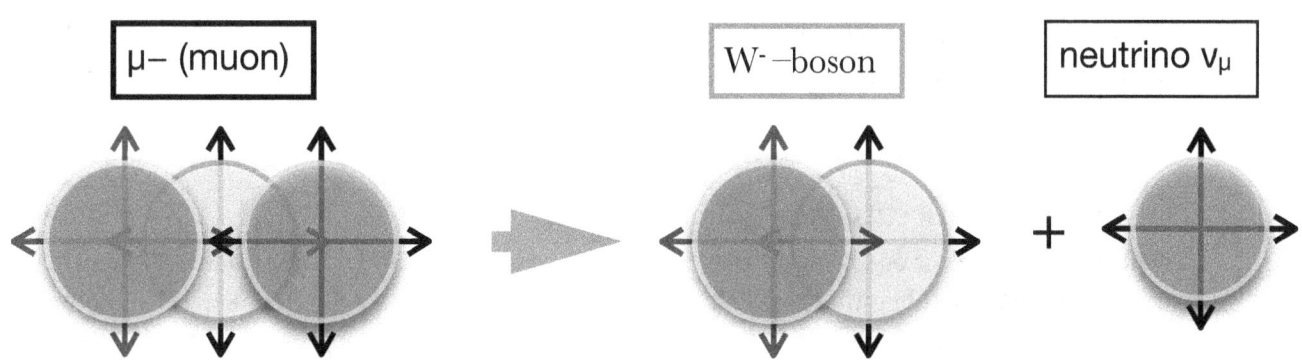

Step 1 Here a negatively charged muon with the spin -1/2 decomposes to a negatively charged W– boson with the spin 0 and a neutrino with charge zero and the spin -1/2. Everything's in order!
Step 2 we recognize from earlier.

Step 2

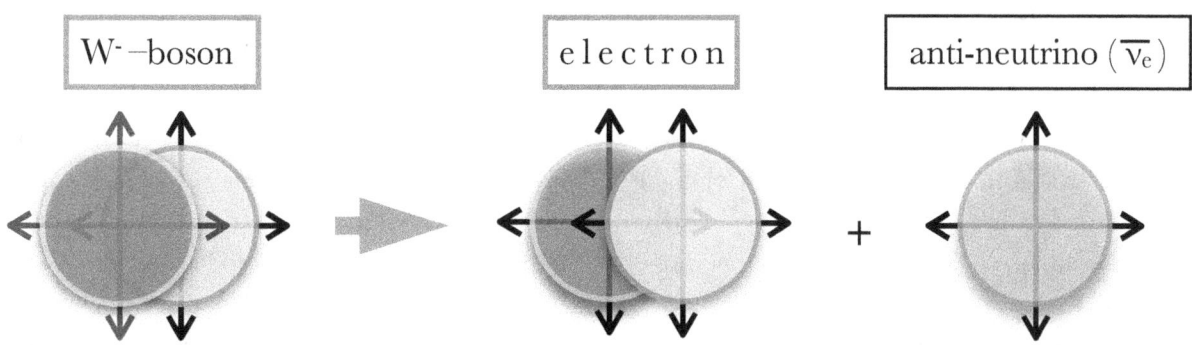

The muon is thus constructed, one can say, of two photons and one neutrino. Or a combination of an electron and a neutrino. Reminds that the division of the photon, thus the formation of a neutrino and an antineutrino occurs according to the known reaction called pair formation.

A combination of a photon and an antifoton gives a W-boson (as in the case of neutron decay) decomposes in an electron and a neutrino.

What distinguishes an electron neutrino (v_e) from a muon dito (v_μ)? The short but true answer is that no one knows. And I will not care about this, a reasonable answer is of course the frequency.

Here's another question. (See: http://www.science20.com/quantum_diaries_survivor/understanding_muon_decay):

How does a muon really decide when and how to decay?

Yesterday somebody asked me here if I could explain how does a muon really decide when and how to decay. I tried to answer this question succintly in the thread, and later realized that my answer, although not perfectly correct in the physics, was actually not devoid of some didactic power. So I decided to recycle it and make it the subject of an independent post.

Before I come to the discussion of how, exactly, does a muon choose when and how to decay, however, let me make a few points about this fascinating particle, by comparing its phenomenology to that of the electron.

Muons versus Electrons

The muon is the electron's heavier brother. It has the same charge and the same status of elementary lepton, but the one thing that makes it different from the electron (well, the other is "muon-ness", which does not count for the sake of this argument) changes most of its phenomenology. Such is the importance of mass!

The electron mass is 0.511 MeV, the muon mass is 105 MeV: a factor of 200 in favor of the latter. Let me make a list below of the difference this makes.

1 - The electron does not have anything lighter to decay into: being the lightest electrically chargedfermion, it is condemned to live forever. The muon, instead, can turn into an electron (indeed I discuss the process in more detail below). The available energy of the reaction is the factor which dictates its lifetime: 2.2 microseconds. So the electron is eternal, while the muon is unstable. The universe would be an entirely different place if muons were as stable as electrons!

2 - When in possess of energy in the range from a MeV to a few GeV, the electron is already quite relativistic (it travels at the speed of light: 297,000 kilometers, or only inches less than that, per second), while at the same energy the muon is much less relativistic. This fact has several important phenomenological consequences, but discussing them would bring me too far today.

3 - The electron radiates bremsstrahlung photons much more readily than muons at the same energy. Take a W boson produced in a LHC collision: if the W decays to an electron-neutrino pair, the energetic electron will have a penchant to radiate out photons whenever it crosses the electromagnetic field of the atoms it traverses.

The muon, with its 200 times higher mass, is subjected to a 1.6 billion times smaller energy loss by radiation at the same energy. This has important consequences for the detection of electrons and muons: the most visible one is that electrons interact in dense matter by producing a shower of secondaries, as I discussed just a few days ago here; muons, on the contrary, pass almost unhindered through large amounts of material. All particle detectors in collider physics experiments are built the way they are because of this simple difference: dense layers of material are used to detect electrons, while muons may be picked up downstream, where no other particles make it.

So how does a muon decay ?

We usually picture an elementary particle as a line in a Feynman graph, and we think of it as a point-like object propagating in space along a straight line. But quantum mechanics tells us otherwise: the particle should rather be pictured as a cloud with undefined boundaries. A lot is going on in its neighborhood: not only does the particle interact with the medium, constantly "talking" with the environment by emitting and absorbing bosons, the quanta of the interacting fields: the particle also emits and reabsorbs "its own stuff", virtual particles it generates by itself. The picture on the right should explain what I mean: the upper straight line is what we may think of the muon propagation; the lower graph shows a more realistic scheme, with the muon continually emitting virtual bosons (dashed lines) which occasionally turn into pairs of fermions, etcetera, ad infinitum.

In particular, the muon emits and reabsorbs virtual W bosons of electric charge equal to its own. By doing that, it momentarily turns into a muon neutrino: electric charge and weak hypercharge get subtracted from it, to be given back as soon as the virtual W comes back from the free trip.

Maybe I should explain what I mean by "virtual" at this junction. A virtual particle is not a ghost, or a mathematical trick! A virtual particle, in truth, is as real as a real one; the only difference is that its presence may be ignored when one describes a particle reaction in the simplest approximation.

Now consider that W boson: however virtual, it is also authorized to temporarily split into fermion-antifermion pairs before it is claimed back by the emitting muonic line. It may thus turn into an electron- electron neutrino pair, for instance; but also other fermion pairs are possible. Then, before you know it, the pair fuses back into the W, and the W fuses back with the muon neutrino to yield the original muon. This is shown graphically on the right.

But with electron-neutrino pairs one different thing may happen: the duo may decide to leave the scene of the virtual fluctuation for good, never fusing back into the W. They do so without violating indefinitely the energy-momentum conservation law, because the total mass of the muon neutrino, the electron, and the electron neutrino is smaller than the original muon mass. So the muon gets fooled: the quantum numbers it lent to the W boson are lost forever. It is effectively turned into a muon neutrino for good.

Muons, as any other subatomic body, constantly emit and reabsorb virtual particles of all the kinds they couple to. The probability that a virtual W plays the disappearance trick to the muon which popped it out of the vacuum is constant every time it gets created: this creates an exponential decay law, no less than the one which applies to the stack of tokens that a unwitty player continues to bet on "red" at the roulette as he wins.

What governs the speed of the disappearance, i.e. the muon lifetime, is a direct consequence of the likelihood of the transition discussed above, which in turn depends on the available energy of the reaction (muon mass minus mass of the final state bodies), as well as the strength of the coupling: the latter is a direct measure of the likelihood of virtual W emission.

Taking Stock

All in all, there is nothing particularly surprising in the description I have provided above: particle reactions readily occur if they do not violate any quantum-mechanical rule. Muons may only decay to electrons -the only particles lighter than muons which carry the same electric charge- and they do so by means of the exchange of a W boson, the carrier of the electroweak charged-current interaction.

Once you understand how muons decay, it becomes absolutely trivial for you to figure out that a b-quark may decay to a charm quark, an electron, and a neutrino: the diagram and the qualitative explanation are the same. But the large mass of the b-quark (4 GeV, 1.2 of which get converted into the mass of the charm quark) also allows different leptons to be produced: a muon-neutrino pair, for instance. A tau-neutrino pair is also possible, but its rate is sizably smaller, since the tau lepton weighs 1.77 GeV, which means it takes away a rather large share of the available energy.

In conclusion, the decay of elementary particles may be understood without the complicated calculations of quantum field theory. A qualitative picture may explain most of the observable features of the process. However, it is quantitative calculations what actually tell us, once compared with experimental measurements, that our qualitative picture is correct!

Why does a muon decompose and not an electron? Well, their components are tied to each other in different ways. The components of the electron are attracting each other first with the force F, until a certain point where they switched. After that, the attractiveness takes over again, but now from the opposite. They quickly take the original position, to revert back to the previous position, etc. The force F is bi-directional. In the muon's case, they attract each other and then repel each other. The restraining force, the one that turns it all around, does not exist. The force F is one-way. As components lose contact and emergence, the photons or neutrionos send back to the imaginary state ...

The forces that hold the electron components.

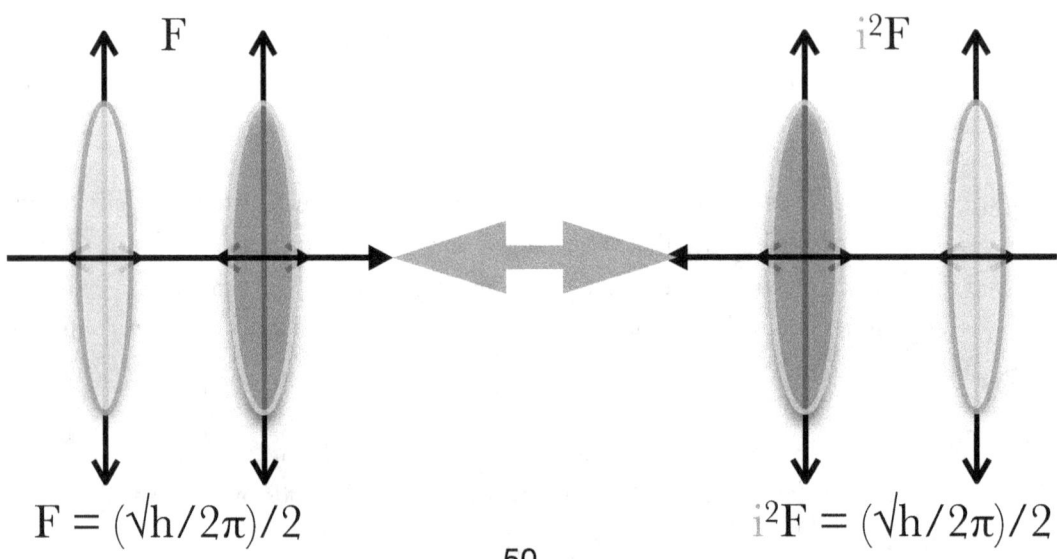

$F = (\sqrt{h/2\pi})/2$ $\qquad\qquad i^2F = (\sqrt{h/2\pi})/2$

What applies to the muon also applies to the W-boson and many other similar particles. Regarding the proton, its triangularly bound neutrinos are a very firm bond, no risk of decomposition. The time of decay is thus linked to Planck's constant h, its constituent power is determining. All of these forces we mentioned are thus electromagnetic nature. There is no particular strong or weak force. Not needed.

"Muons, as any other subatomic body, constantly emit and reabsorb virtual particles of all the kinds they couple to", is stated in the quotation above. Again: Not needed.

"But the large mass of the b-quark (4 GeV, 1.2 of which get converted into the mass of the charm quark) also allows different leptons to be produced: a muon-neutrino pair, for instance." Once again: Not needed.

As for the electron's other relative the tauon, the reader can now readily analyze the composition of. And give it a structure. Same with the pion. The next step in the analysis of particle physics is now the atom. But it requires a special chapter, which I will return to ...

***

E-mail:

akehedberg2015@gmail.com

akehedberg@kiruna.nu

www.sofia.linnea.com/~akejean/

Universe's origin and way of working

kosmos2

YouTube:

Cosmos and Universe: How does it work without a Big Bang?

https://youtu.be/D4SjZOkV2Z4

https://www.youtube.com/edit?video_id=D4SjZOkV2Z4&video_referrer=watch

Books and writings on natural science published by Åke Hedberg, available on many online bookshops:

Hur, när och varifrån fick DNA-molekylen sin programkod?

Förslag till katalytisk Fusions-Reaktor (i stället för ITER, TOKAMAK m.fl.)

'What are light quanta?'

Instead of the ITER project and the TOKAMAK principle Universums byggstenar

Universum utan Big Bang & Atomer utan kvarkar.

Postscript

The superficial philosophy that believes that the electron lacks content and thus structure, and thus is punctual, has its roots that can be traced back to the 1920s, as I see it.

Then founded the positivist philosophy we find in today's physics and other natural sciences. The contents will be destroyed in Heisenberg's (and Bohr's) philosophy, the so-called Copenhagen interpretation where the statistical data and mathematical formulas is all we can lean on.

A philosophy Albert Einstein stubbornly criticized his entire life, as we know. Only fuzzy statistics that "God plays dice", he meant. This is often the content of any criticism of the Copenhagen interpretation, and it was Einstein's. The crisis, the physicist and the famous philosopher Karl Popper saw in the new physics, in his opinion, was essentially due to two things: a) the interference and the penetration of subjectivism in physics; and b) that idea conjured as saying that quantum theory has reached a complete and final truth.

The belief that quantum mechanics is the final truth is a major obstacle to scientific development. Therefore Einstein could not convince Bohr and others with his criticism. There were also some important puzzles missing for this at this time. More sophisticated instruments were required than were available in the 1920s and 1930s, especially computers. Therefore, it took a long time in the 1980s before an advanced chaos theory could be established.

But the established truth about the perfect quantum mechanics prevented an integration with new discoveries and theories. Thus, one has not realized that nature was and is organized not only in the mechanical way that quantum **mechanics** and wave **mechanics** assume. What chaos theory using fast computers so graphically and vividly could show.

It is thus no coincidence that physics deals with punctual mechanical objects, but is a consequence of a world view, an ideology. In fact, there is nothing in this world that does not have a structure. We have seen that even the photon and neutrino are composite objects.

***

www.ingramcontent.com/pod-product-compliance
Lightning Source LLC
Chambersburg PA
CBHW081815220526
45470CB00006B/2319